Sprinkler Irrigation and Agricultural Development in Drought Prone Ecosystem

About the Author

Professor Noor Mohammad (b.1942) after obtaining his M.A. Degree in Geography from the University of Gorakhpur with a first class first, joined Aligarh Muslim University, Aligarh in 1965 as a Junior Research Fellow. He was awarded Degree of Ph.D. in 1969 and joined Kirori Mal College, Delhi in the same year. Since 1976 Prof. Mohammad has been associated with the Delhi School of Economics, Delhi University, in various capacities, such as, Assistant Prof., Associate Prof., Prof. Head of Department, Co-ordinator, UGC SAP Programme. He has taught Geography from graduate to Ph.D. scholars for about four decades.

He has been also associated in various positions and capacities with many institutions and organisations, such as, UGC, ICSSR, NCERT, CBSE, NIOS etc. He has completed many Research Projects of ICSSR, UGC, NCERT. Prof. Mohammad has been awarded UGC Emeritus Fellowship and has completed the latest Research Report on *Impact of Sprinkler Irrigation on Agricultural Development In Drought Prone Areas of Haryana.*

Prof. Mohammad is actively engaged in research in Applied Geography and Agricultural Geography and has more than hundred research papers to his credit in many scholarly referred scientific journals. He has authored and edited more than two dozen books.

Sprinkler Irrigation and Agricultural Development in Drought Prone Ecosystem

A CASE STUDY OF HARYANA

Noor Mohammad
Emeritus Fellow
Department of Geography,
Delhi School of Economics
University of Delhi
Delhi-110 007

CONCEPT PUBLISHING COMPANY PVT. LTD.
NEW DELHI-110059

ISBN-13: 978-93-5125-224-5

First Published 2016

Published and Printed by:

Concept Since 1974

Concept Publishing Company Pvt. Ltd.
A/15-16, Commercial Block, Mohan Garden,
New Delhi-110059 (INDIA)
Phones : 25351460, 25351794, *Fax* : 091-11-25357109
E-mail : publishing@conceptpub.com
Website: www.conceptpub.com

Editorial Office: H-13, Bali Nagar, New Delhi-110 015, India.

Cataloging in Publication Data—DK Courtesy: D.K. Agencies (P) Ltd. <docinfo@dkagencies.com>

Mohammad, Noor, 1942- **author.**
Sprinkler irrigation and agricultural development in drought prone ecosystem : a case study of Haryana / Noor Mohammad.
pages cm
Includes bibliographical references and index.
ISBN 9789351252245

1. Sprinkler irrigation—India—Haryana. 2. Land use, Rural—India—Haryana. 3. Crops—Effect of drought on—India—Haryana. 4. Arid regions agriculture—India—Haryana. I. Title.

S619.S66M64 2016 DDC 631.5870954558 23

Preface

Water is prime input and foundation of agriculture whether it is available through precipitation or irrigation because even the metabolic process will not take place without it what to say of vigorous vegetative growth of plants and higher yield of crops. But the rainfall which is the prime source of water, is periodic, seasonal, and highly variable in its time of incidence, retreat, amount and its intensity. All these uncertainties in the occurrence of precipitation necessitate the use of irrigation water. Therefore, irrigation facilities are decisive factors in crop regionalization than any other determinant in the inherently fertile dry lands. Irrigation being an artery and pulsating heart, is also an absolute constraint as well as sufficient volitional command over the location of commercial crops, improvement in the yield per hectare, swing in cropping pattern and dynamism in livestock structure. Therefore, the availability of adequate, timely and assured supply of water is an important determinant of agricultural productivity.

Although India is blessed with abundant water resources, but the water available for irrigation purposes has been continuously diminishing. The use of water for irrigation constitutes about 84 per cent of total utilised potential water, i.e. 75 billion cubic metres. With the increase in demand of water for other uses, the share of water use for irrigation is likely to go down to nearly 73 per cent by the year 2025.

In Haryana, a water thirsty state, irrigation and power, by and large, have been given top priority in the development plans since its inception so as to develop the agriculture at faster rate. Therefore, Haryana has made commendable progress through irrigation facilities in laying the foundation of progressive and forward looking agricultural activities. In the recent past with the induction of minor irrigation development schemes such as sprinkler and drip irrigation, the irrigation facilities have considerably enhanced.

However, there are considerable regional imbalances in the irrigation facilities and spatial differences in the trends of irrigation development that have swerved from the principle of equality. With areal differences in topography, water table, quality of ground water etc. in Haryana, the sources of irrigation, irrigation facilities, and intensity of irrigation would naturally not be uniform both in space and time. And, it is more so in the drought prone areas of Haryana. Therefore, there is a need to study the impact of sprinkler irrigation on agricultural development.

The study is based on primary and secondary data. The secondary data have been collected from published and unpublished records of the Government and Semi-Government organizations of Haryana and also from the NGOs relevant websites of Haryana. Primary data have been collected for 465 households chosen from 13 selected villages lying in different development blocks of the state. The data have been processed and analysed using appropriate statistical techniques and have been presented in tabular, diagramatic, charts and map forms using GIS software.

The study is organized in 9 chapters. It opens up with an introductory chapter dealing with conceptual framework, statement of problems, research questions, aim and objectives, hypotheses, study area and overview of literature based on themes. Chapter 2 deals with research design highlighting in detail the procedure of primary data collection, data base and methods of data processing, analysis and presentation, while Chapter 3 discusses the methods of identification and delineation of Drought Prone Areas of Haryana. The geographical profile of the study area has been presented in Chapter 4 which includes physical, economic and social setting of the area. In this chapter only those aspects of geographical environment have been taken which are interlinked with one another and have a bearing on the agricultural development in the state. The sources and pattern of irrigation specially the sprinkler irrigation have been described and explained in chapter 5.

Chapter 6 discusses the spatial pattern of land use/land cover with emphasis on intensity of cropping in 2005-06. All the five major categories of land use have been systematically analysed focusing on those aspects which could help in agricultural diversification and development.

Chapters 7 and 8 are the core chapters of the study. Chapter 7 represents the spatial pattern of agricultural land use and level of its development. It describes and explains the spatial distribution of major crops such as *bajra*, wheat, pulses, oilseeds etc. It discusses the crop combination regions and levels of agricultural development using appropriate statistical techniques. Chapter 8 describes and explains the socio-economic transformation of the people using 6 social and 5 economic indicators. Composite indices of social, economic, and socio-economic transformation have been worked out in order to examine the transformation in them.

Based on the critical analysis of the facts and figures in the previous chapters, summary, conclusion and suggestions have been given in last chapter which derives innovative conclusions and gives pragmatic and practical suggestions.

Delhi
25th August, 2016

Noor Mohammad
Emeritus Fellow
Delhi University, Delhi

Acknowledgements

Let me first of all thank the Almighty, Whose grace and mercy has always been with me and provided all I needed which I myself was not aware of.

The completion of this project has been possible due to the constant inspiration and unstinting co-operation of a host of people too numerous to acknowledge individually. However, first and the foremost I express my deep sense of indebtedness and gratitude to my *Guru* and mentor late Prof. Mohammad Shafi. Professor Emeritus, Department of Geography, Aligarh Muslim University, Aligarh, who has been a source of inspiration and guidance throughout my academic career. Whatever I have achieved in my academic pursuits, is due to his persistent and sincere able guidance.

I owe a deep sense of gratitude to my late parents who have shaped and made me what I am today. It is their soul and spirit which is still inspiring and giving guidance and strength in every walk of my life. I express my sincere thanks to University Grants Commission especially to its former officiating Secretary Dr. R.K. Chauhan who has been instrumental in the award of the Emeritus Fellowship to me.

I extend my thanks to the Head, Department of Geography, Delhi University who has kindly permitted me to use the infrastructural facilities of the department. I would also like to extend my thanks to all the Faculty members of the department who have helped me in one way or the other in the completion of this work.

I have consulted many libraries within and outside Delhi specially the libraries of Delhi University, Jawaharlal Nehru University, Jamia Millia Islamia, British Council, ICSSR, NCERT, Kumaon University (Nainital) and Garhwal University (Srinagar). I extend my heartiest thanks to all the librarians and their staffs who have been kind enough to allow me to use the relevant data and existing valuable literature in these libraries.

The entire fieldwork and collection of primary and secondary data have been done by fellow assistant who has collected very valuable primary and secondary data. I am also grateful to respondents of all selected villages, village heads, and the officials of districts, tahsils and block development offices. Needless to mention that without their help and co-operation field investigation would have been very difficult. It is the lack of space rather than my sense of gratitude which bars me from naming them individually.

I would also like to express my thanks to M/s Arun Thesis for typing the manuscript and bringing it in the present form. Thanks are also due to Dr. Perveen Misra, Miranda House, Delhi University, for drawing all the maps and diagrams within a short period of time.

I would like to put on record my sincere thanks and deep sense of appreciation to my wife Mrs. S. Nisa and children—Nuzhat, Naved, Zareena and Javed who have provided a favourable working environment to me for successful completion of this research project. Last but not the least, I would like to put a special word of appreciation, love and affection to my grandchildren—Eram, Zaheen, Uzair, Suhema and Affan who have been a source of encouragement and inspiration. I would like to add a special word of appreciation to beloved Affan who use to cheer, inspire, and refresh me whenever I felt tired and trophic level of mind gets jammed ın writing this book.

Noor Mohammad
Emeritus Fellow

Contents

List of Tables

List of Maps

1
Introduction

1.1 Introduction

Water is the essence of life on which man depends not only for his direct consumption needs, but also for his food and fibre. That is why early civilization took place by the sides of the rivers. Irrigation is as old as civilization. No doubt irrigation has been lifeline of Indian agriculture since time immemorial, but it assumed greater importance after the induction of "New Agricultural Strategy" in 1966-67 in the name of Green Revolution. The additional requirements of food and fibre to meet the demands of growing population have been realised to be met by expansion of agriculture to marginal areas such as dry land areas which normally face droughts and famine. It is reported that about 260 million hectares are subjected to different degree of water stress and drought conditions in India (Nagrajan, 2003). Annually about 19 per cent geographical area and 16 per cent population of India are vulnerable to drought disasters because a large number of people and their livelihood are dependent on monsoon (Singh, 1998). This includes 38.7 million hectares of arid areas, of which 32 million hectares of hot arid region is found in the state of Rajasthan (19.52 m.ha), Gujarat (6.40 m.ha), south-western part of Punjab and Haryana (2.88 m.ha) and certain areas of Maharashtra, Andhra Pradesh and

Tamil Nadu (together 3.20 m.ha). In Haryana, Bhiwani, Mahendragarh and parts of Hisar and Gurgaon districts are drought-prone areas. Drought-prone areas have various social, economic and environmental problems of which some are unavoidable, while other can be planned out. Such arid regions are characterized by sandy undulating topography, high temperature, low rainfall, scanty or almost absence of vegetation. With the help of irrigation, some of dry land areas can be brought under cultivation.

Irrigation is generally defined as the artificial supply of water to soil for the purpose of supplying the moisture essential for plant growth. It plays a paramount role in increasing the yield, agricultural output, and enhancing cropping intensity and agricultural productivity (Dhawan, 1988 and Vaidyanathan *et al.*, 1994). Besides benefiting the farmers, irrigation development also helps to increase employment opportunities and wage rate of the agricultural landless labourers which are essential for reducing the poverty among the landless labour households (Bhattarai and Narayanmoorthy, 2003; Narayanmoorthy and Deshpande, 2003; and Saleth, 2004). But water is becoming increasingly scarce worldwide due to various reasons specially fast expansion of population and economic activity (Rosegrant *et al.*, 2002). The problem of water scarcity is expected to be aggravated further. Micro-level estimate indicates that one-third of the word population would face absolute water scarcity by the year 2025 and the worst affected areas would be the semi-arid regions of Asia, the Middle East and Sub-Saharan Africa (Seckler *et al.*, 1998 and 1999). India too has started facing severe water scarcity in different regions at a faster rate (Saleth, 1996 and CWC, 1905). The agricultural sector which currently consumes over 80 per cent of the available water in India continues to be major water consuming sector due to its intensification (Saleth, 1996, and Iyer, 2003). One of the main reasons for the low coverage of irrigation is the predominant use of flood method of irrigation in which water use efficiency is only 35 to 40 per cent because of huge conveyance, seepage and distribution losses (Rosegrant, 1997).

Considering the water availability for future use and the increasing demand of water from different sectors, a number of demand management strategies and programmes have been introduced since late seventies in India to increase water efficiency use. One of the demand strategies introduced relatively recently to control water consumption in Indian agriculture is micro-irrigation which includes mainly drip and sprinkler irrigation method in which water is supplied at a required interval and quantity using pipe network, emitters and nozzles. Therefore, the conveyance and distribution losses are reduced almost completely which results in higher water use efficiency. Sprinkler irrigation method sprinkles water similar to rainfall in the air through nozzles which subsequently breaks into the small water drops and fall on the field surface, whole in drip irrigation water is supplied to the root of the plants at required interval. The on-farm irrigation efficiency of properly designed and managed drip irrigation system is estimated to be about 90 per cent, while the same is about 35 to 40 per cent for surface method of irrigation. In case of sprinkler irrigation method it is about 70 per cent as it supplies water over the entire field of crop (Kulkarni, 2005). The productively gain due to use of sprinkler irrigation is estimated to be 20-70 per cent for different crops. It also reduces weed problem, soil erosion, energy use and cost of cultivation substantially (Narayanmoorthy, 1995 and 2001). It may also be adopted for all kinds of land surface which is generally not possible through flood method of irrigation. Sprinkler irrigation method is considered to be the safest method for uneven and sandy soils. The system consists of sprinkle heads or nozzles which are mounted on rise in lateral lines taken from main line which is further connected with pumping unit. The system is suitable for almost all crops except very water intensive crops like rice and sugarcane. It is well suited for the sandy soil with excessive percolation and undulating land and slopes too steep to level, and water scarce areas. It increases irrigated area by 40 per cent with the same amount of water, about 40-60 per cent labour is saved and the wastage of land in making bunds and channels is also avoided. It also increases the gross cropped area by bringing the land devoted to irrigation channels, embankments, bund, etc. under crop cultivation. Nearly 15-20 per cent area increases in this method of irrigation (Kumar, 2011). According to the latest studies the agricultural plants absorb only 25-30 per cent of irrigation water in the surface irrigation method and remaining 70-75 per cent goes waste either by evaporating or by percolating down. As such sprinkler irrigation conserves the most valuable resource, i.e. water. Such an advantageous system is the only alternative for managing the water in drought-prone areas. It is reported that about 658,500 hectares of land was under sprinkler irrigation in India in 1997-98. The area under this system was the highest in Madhya Pradesh followed by West Bengal, Assam and Haryana but during 2003-04 the areal strength of sprinkler irrigation has increased to 16,34,997 hectares with Haryana (29.97%) as a leading state followed by Rajasthan (25.99%) and West Bengal (8.26%). It shows that this system is increasing day-by-day in Haryana. Hence, the present study deals with this newly adopted system of irrigation and its role in agricultural development in drought-prone areas of Haryana. It will also assess its impact on socio-economic structure as a result of increasing level of prosperity due to agricultural development.

1.2 Statement of Research Problem

The State of Haryana has been carved out from erstwhile Punjab

State on November 1, 1966. Covering an area of 404 million hectares, it lies between 27° 39′ to 30° 55′N. Lat. and 74° 27′ 8″ to 77° 36′5′ E. Long. (Fig. 1.1). Here out of 3.9 million hectares of net cultivated area, irrigation facilities were available only for 1.29 million hectares in 1966-67. More than 65 per cent area of state lies in arid and semi-arid belt. So, net irrigated area is very less particularly in south-western Haryana, where land is too uneven and dotted with permanent sand dunes. There are wide differences in output and productivity in the state and its districts. Northern region comprising Ambala and Karnal, is at the top, central region consisting of Rohtak, Sonipat and Karnal stands in the middle and south-western region comprising Bhiwani, Hisar, Mahendragarh and Rewari is at the bottom. These differences are mainly due to differential uses of crucial inputs of irrigation, HYV seeds and fertilizers.

Even after taking various measures with conventional methods of irrigation, the situation is still grave in drought-prone areas of the state. This part of Haryana bordering Rajasthan demands special water management approach because of light textured soils and moderate to highly undulating topography. The groundwater which is marginal in quality is not only brackish, but also found at greater depths with less volume of water. The other contrary problem of the region is that of rapidly rising water table in some pockets due to negligible withdrawal coupled with continuous recharging of groundwater regime by canal water. Keeping all these factors in view, the sprinkler irrigation which increases irrigation efficiency has been considered as a viable alternative method of irrigation in drought-prone areas of Haryana.

During the year of 1969, about 40 sprinkler sets were purchased and installed under Desert Development Scheme for demonstration purpose by Irrigation Department of Haryana. The Agriculture Department, Haryana introduced a "Sprinkler Irrigation Scheme" in the state during 1971-72. The number of sets increased to 10407 in 1983, 65291 in 1996 and 99845 in 2003 in the state. But still this system of irrigation is used only in south-western parts of Haryana. Even in this part also only the richer landlords have used this. So, State Government has now subsidized it by 25 per cent, so that, this system can go to poor farmers also. With its adoption, the farmers have raised their living standards as their agricultural development has taken place. Now, this system of irrigation has become very popular in the area. Consequently, the agriculture has considerably developed and has brought economic prosperity to farmers which in turn has been instrumental in bringing socio-economic transformation in the region. But it is not uniform across the region which needs to be thoroughly researched and investigated at meso- and micro-levels.

1.3 Research Questions

In this given scenario, as discussed above, the following research questions arise:

(i) How far the sprinkler irrigation is viable in this area?
(ii) How much sprinkler irrigation is being adopted by the farmers?
(iii) How far sprinkler irrigation will solve the problem of water scarcity of undulating topography of drought-prone areas?
(iv) How far will the sprinkler irrigation be helpful in raising the cultivated area and agricultural production to feed the growing population?
(v) How far sprinkler irrigation has changed the scenario of the drought-prone areas so that the same can be applied to the other dry land areas also?

The present study intends to research and investigate some of these issues in detail at macro and micro levels.

1.4 Aims and Objectives

The basic aims and objectives of the study are:

(i) To identify and delineate drought prone-areas of Haryana.
(ii) To trace out the spatial pattern of distribution of sprinkler irrigation.
(iii) To assess the impact of sprinkler irrigation on cropping pattern and agricultural productivity.
(iv) To examine the level of agricultural development and its impact on socio-economic conditions.
(v) To suggest a viable strategy for application of sprinkler irrigation.

1.5 Hypotheses

(i) Sprinkler irrigation has positive correlation with agricultural development.
(ii) Agricultural development leads to socio-economic transformation.

1.6 The Study Area

The present study has been carried out in the chronically drought-prone areas of South-western Haryana. Covering an area of 6429 square kilometres it lies between 27° 42′ to 29° 15′ N. Lat., and 75° 15′10″ to 76° 30′15″ E. Long. (Fig. 1.1). It is in the form of an elongated belt bordering Rajasthan. It includes Hisar-II and southern part of Hisar-I block of Hisar district, Siwani, Tosham,

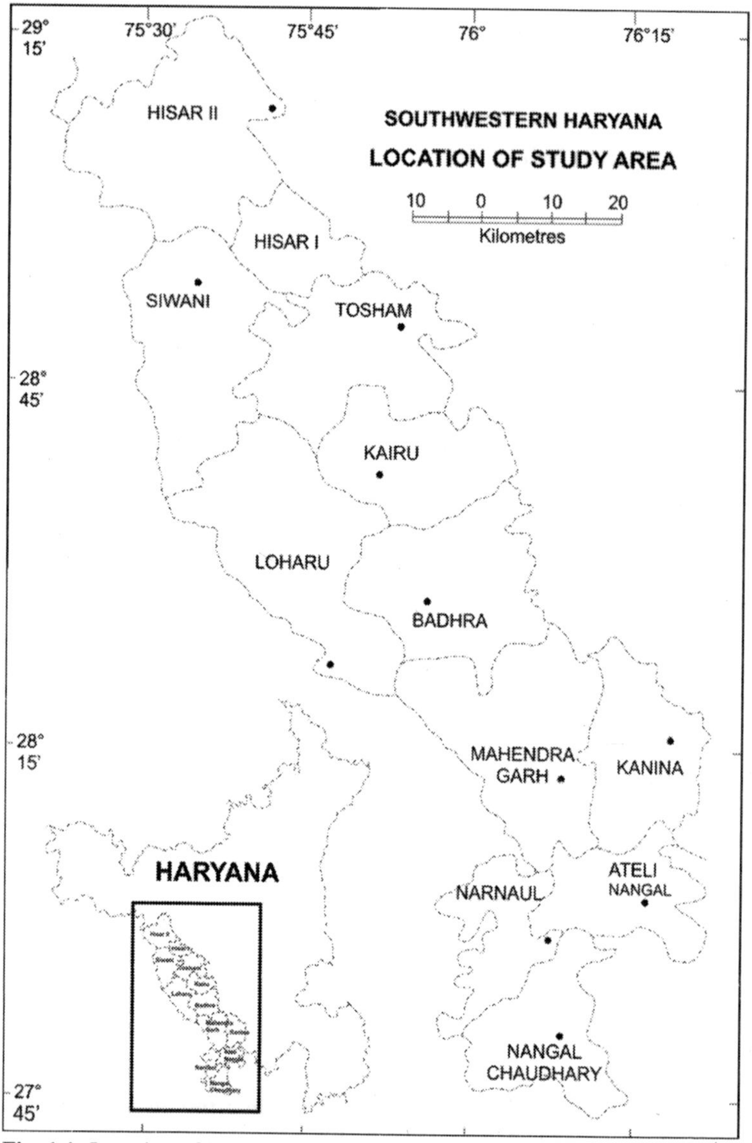

Fig. 1.1: Location of Study Area

Loharu, Kairu, Badhra and parts of Bhiwani and Dadri blocks of Bhiwani district, and Kanina, Mahendragarh, Narnaul, Ateli Nangal and Nangal Chandhry blocks of Mahendragarh districts of Haryana. It is bounded by Fatehabad district in north, Rohtak, Jhajhar and Rewari districts of Haryana in east, Churu district of Rajasthan in south and Jhunjhunu district of Rajasthan in the west.

The sandy undulating topography of the belt slopes from south-west to north-east. The region is dotted with sand dunes of various shapes and sizes which form a thirsty land covered with steppe vegetation. Mostly the soils are sandy and loamy sand (*Bangar*), sandy loam and fine loamy. Rocky surfaces are also found in the area which are offshoots of the Aravali and they stand out distinctly against sandy swelling waves.

The climate of the region is characterized by the dryness, except during the monsoon which lasts during July to August. The temperature reaches as high as 40°C during May and as low as 2°C during January. The very scanty annual rainfall in the belt varies from 30-40 cms and in some areas it is even less than 30 cms. The average annual potential evapotranspiration is about 145 cms. So, the average moisture balance is in deficit.

The economy of the area is primarily agricultural. The agricultural year is broadly divided in two seasons: *Kharif* and *Rabi* which are locally known as '*Swain*' and '*Sadri*' respectively. The major *kharif* crops are *bajra*, *jowar* and *moong* (pulse), while the dominant *rabi* crops are wheat, gram, barley and *sarson*. The modern multi-stage lift irrigation schemes such as Jawaharlal Nehru, Jui, Loharu, Siwani lift irrigation schemes, have come-up which have helped in horizontal expansion of agricultural land and its productivity.

Deficiency and fluctuations in rainfall are serious problem in the region because the average rainfall is not sufficient to meet the needs of even the dry food crops. The underground water in almost whole of the tract is brackish and found at a greater depth (6 m to 50 m below the surface). The level of this area rises towards the south with the result that the water from gravity flow through canals, which all start from the north cannot be automatically utilized to irrigate thirsty land. This chronic shortage of water necessitates maximum and optimum use of irrigation water wherever and whenever it is available. It is, therefore, important to adopt sprinkler irrigation in the study area for the judicious use of the scarce water resource which in turn will increase agricultural production from the existing land.

Due to these variable characteristics, this particular drought-prone region has been chosen as the study area for the present study.

1.7 Review of Literature

From the detailed literature survey conducted for the present theme of research, it has been found that there are lot of general studies related to irrigation and agricultural development, but the studies directly related to sprinkler irrigation in drought-prone areas are very few. Literature survey has been conducted on three sub-topics, i.e. irrigation, sprinkler irrigation and its impact on agriculture development in drought-prone areas. A few relevant studies for the present work are presented here.

Dagal (1997) observes that drought is complex and multi-faceted phenomenon. The departure from normal precipitation provides only one dimensional definition of drought. People's perceptions of rainfall deficiency, agricultural operation, affected scarcity of food, drinking water and fodder, timing of rainfall can help us to understand several dimensions of drought.

Sen (2000) finds that drought may adversely affect food production and agricultural expansion but famine and hunger relate not only to food production and agricultural expansion, but also to the functioning of the entire economy and even more broadly the operations of the political and social arrangements.

Dash and Behura (2000) on the basis of primary and secondary data collected from chronically drought affected villages in Kalahandi, have pointed out that drought situation in Kalahandi was outcome of both— 'natural and anthropogenic factors'. Failure of rainfall as natural and deforestation as human made factor, have almost equally contributed to drought situation.

Simhadri (1997) in his paper refers to grim situation of drought- prone Telangana region, despite of two rivers are traversing this region. This situation has emerged on account of two man-made reasons. One is people in this region resort unsustainable agricultural practices, such as growing water intensive crops. Another reason is simply taking water out of two mighty rivers for the agricultural and industrial development of coastal region at the cost of drought-prone Telangana. Desai (2003) discusses the impact of drought by considering how agriculture contributes to economic development. He also discusses how to mitigate the adverse consequences of drought both in short- and long-run. Desai pointed out that drought not only affects agriculture adversely, but whole country. Singh (1998) in his paper aims at stressing the need to evolve appropriate drought reduction strategies keeping in mind the importance of protecting people of drought-prone area from the impact of rainfall fluctuation and monsoonal uncertainties.

Planning Commission (1973) had set-up a taskforce under the chairmanship of B.S. Minhas which made first systematic effort to provide a long term perspective for the integrated agricultural development in drought-prone areas. The task force points out that although all drought-prone areas experience erratic rainfall but they may have same water and soil situation. Government of India

(1994) constituted a committee which conducted extensive study of drought-prone areas and found that the core objective of various programmes to restore ecological balance through soil and moisture conservation on water shed, has been widely neglected. These programmes are being implemented on *ad hoc* basis in a fragmented manner. Karanth (1991) reveals some other strategies adopted in Karnataka village such as growing a variety of crops (mixed crops) and rearing livestock, entering labour and tenancy, drawing upon the previous stock of grain of fixed assets, adjusting consumption, borrowing and relaying upon traditional social security system.

Sharma and Jain (1981) have made an attempt to analyse the spatio-temporal variations in irrigation in order to see their correlation with the levels of agricultural development in Madhya Pradesh. An appraisal of present irrigated farming system, its changes over time and space and scope for future development in Rajasthan desert has been made by Ram and Sen (1981) in their article. Kumar and Pal (1985) in their article have discussed that a drastically improved and sophisticated irrigation has contributed to the shift of Haryana's agriculture. Singh (1992) has discussed the impact of different sources of irrigation on cropping pattern, yields, and farm practices in Aligarh district of UP. Singal (1995) has discussed the remarkable transformation in the socio-economic conditions brought by adopting various policies in water resource development in arid areas of Haryana in his article. Sandra (1999) examined the rise and fall of irrigation based civilizations explaining impact of irrigation on society. Goswami *et al*. (2002) have assessed the impact of irrigation on land use efficiency and area allocation in different size classes of land holding in Maharashtra. The result of this study have indicated that farmers diverted their irrigated areas towards cultivation of cash crops for better financial return. Sengupta (2002) has made an attempt to study the water resources of Malda district. The carrying capacity of land in terms of SNU has been estimated and on this basis, food surplus and deficit areas have been identified.

Laxminarayan (1984) has discussed the ex-post evaluation of investment on sprinkler irrigation in Haryana at the instance of Agricultural Reform and Development Corporation. Aggarwal (1985) in his article has discussed sprinkler irrigation and its comparison with other methods of irrigation. He has also examined the economic feasibility of sprinkler irrigation in desert areas of Haryana. Sivanappan (1987) has discussed the conceptual framework related to sprinkler irrigation in India, importance of Sprinkler irrigation, types of sprinkler system etc. in his book. Hooda (1990) has tried to study the role of sprinkler irrigation in the transformation of agriculture in the arid areas of Bhiwani district of Haryana. Singh and Aggarwal (1993) have discussed the adoptability and advantages of sprinkler system in Haryana. They focused on the proper design of the sprinkler system in their article. Kashyap and Lohan (1995) have assessed the impacts of "sprinkler irrigation scheme" started in 1971-72 in Haryana state so as to know-how the farmers are benefited with this scheme and what is the extent of agricultural production in the state after adopting it. Knox *et al.* (2000) have tried to find out financial impacts of sprinkler irrigation. They have traced out the financial benefits of sprinkler irrigation with the help of various maps in their article. Tarjuelo *et al.* (2000) has discussed the changes in evaporation and drift losses in irrigation with change in medium size sprinklers. They have studied this impact under semi-arid region. Mclean, *et al.* (2000) have made an attempt to trace out the spray evaporation losses from sprinkler irrigation system. Ball and Sammis (2001), in their article, have compared the irrigation efficiencies of sprinkler, trickle and furrow for a particular crop of onion production. Li and Rao (2001) have discussed the impact of sprinkler irrigation system on the crop yield in general.

Similarly, there are number of studies dealing with various components of development. Bedi (1982) has tried to find out the post-Independence transformation of rural landscape of central Haryana. More and Mustafa (1984) have studied the irrigation requirements and developments in Maharashtra in their article. The study conducted by Abdul (1988) reveals that change in agricultural productivity has affected the regional development in sub-Himalayan east region of Uttar Pradesh. Shafi (1990) has made an attempt to study agricultural and rural transformation in India. Abbunoori, Abbas Ali (1990) has tried to explain the relation between the irrigation and agricultural development in Drought-prone areas of Iran. Kazi and Wodeyar (2001) have made an attempt to focus on agricultural situation, development and expected changes in cropping pattern and farming systems in the north-eastern region of Bangladesh. Sangwan (2002) has made an study of the sample villages from western Haryana (Sirsa and Bhiwani districts). Nagrajan (2003) in his book has explained drought, its assessment, monitoring, management and resource conservation.

Some of the studies (Dhawan, 2002) suggest that yield improvement and water saving in sprinkler irrigation is less striking, though the picture vary considerably across crops and space. The result of experimental report pertaining to 15 crops (foodgrains, oilseeds and other crops) shows considerable amount of water saving and yield improvement in all crops. The water saving due to sprinkler irrigation is relatively higher among foodgrains crops, whereas relatively higher yield gain is observed in oilseed crops.

A large number of individual evaluation studies on sprinkler irrigation carried out in different crops have also revealed that water saving and yield improvement is comparatively more in it than other methods of irrigation.

Similarly, another experimental result of three years trial conducted in Gujarat with winter maize indicates the superiority of sprinkler irrigation over flood irrigation method. The trial shows about 40 per cent of water saving and 30 per cent of productivity gain over the conventional method of irrigation (Patel *et al.*, 1993).

In order to find out the relative economics of sprinkler irrigation over flood irrigation method in certain crops, a three-year trial was conducted in north Gujarat. The study shows that under the conditions of normal water availability, the use of sprinkler irrigation method can result in 29 per cent water saving along with 35 per cent yield increase. Importantly for farmers, the net income per mm of water used rose from Rs. 2.80 with surface method to Rs. 9.57 with sprinkler method of irrigation (Mehta *et al.*, 1993).

A yield experiment conducted in south Gujarat during 1989 to 1991 to explore the feasibility of adoption of mini-sprinkler irrigation for onion crop during *rabi* season shows a considerable increase of yield and water saving over the method of flood irrigation (Desai *et al.*, 1993). Similar to onion, another experimental study carried out in south Gujarat to study the feasibility of mini-sprinkler in sunflower also shows that its gain in increasing yield (about 13%), water saving (20.31%) and water use efficiency (47.61%) as compared to conventional method of irrigation (Patel *et al.*, 1993). Similar results were also reported by Abrol and Sharma (1990) in sunflower crop.

The overview of literature reveals that there is little work on impact of sprinkler irrigation on agricultural development leading to socio-economic transformation. Therefore, the present study has been taken up to assess the impact of sprinkler irrigation on agricultural development and its effect on socio-economic conditions in drought-prone areas of Haryana.

References

Aggarwal, M.C., 1985: Sprinkler Irrigation: A Case Study in Haryana, *Indian Agricultural Research Institute*, New Delhi.

Agrews, C., 2002: Drought, Desertification, and Desiccation: Need for Further Analysis, *Geography*, Vol. 87, No. 3.

Bhattarai, N. and Narayanmoorthy, 2003: Impact of Irrigation on Rural Poverty in India: An Aggregate Panel Data Analysis, *Water Policy*, Vol. 5, Nos. 5-6.

Butola, B.S., 2003: Drought of Doom: Its story of Rajasthan, *Journal of Geography*, Vol. 3, No. 5.

CWC, 1905: Water and Related Statistics, *Central Water Commission*. Government of India, New Delhi.

Dagal, K.C., 1997: Defining Droughts in Marginal Areas: The role of Perception, *Professional Geographer*, Vol. 49, No. 2.

Desai, B.M., 2003: Drought Impact and Vision of Proofing, *EPW*, Vol. 38, No. 21.

Desai, D.N., *et al.*, 1993: *Feasibility of Mini-sprinkler in Onion*, Central Board of Irrigation and Power (CBIP).

Dhawan, B.D, 1988: Role of Irrigation in Raising Intensity of Cropping, *Journal of Indian School of Political Economics*, Vol. 3, No. 4.

Dhawan, B.D., 2002: *Technological Change in Indian Irrigated Agriculture: A Study of Water Saving Methods*, Commonwealth Publishers, New Delhi.

Gohring, T.R. and Wallender, W.W., 1987: Economics of Sprinkler Irrigation System, *Agricultural Economics Journal of American Society*, No. 30, No. 4.

Goswani, S.N. *et al.*, 2002: Impact of Irrigation on Landuse Efficiency and Area Allocation in Different Class Size of Landholdings in Maharashtra, *Geographical Review of India*, Vol. 64, No. 4. Kolkata.

Hooda, R.S., 1996: Diffusion of Sprinkler Irrigation in Arid India, *Crop Research*, Vol. XI, No. 2.

Hooda, R.S., 1990: Sprinkler Irrigation and Agricultural Transformation in Bhiwani District, Haryana, *Geographical Review of India*, Vol. 52, No. 3, Kolkata.

Iyer, R.R., 2003: *Water: Perspectives, Issues, Concern*, Sage Publication, New Delhi.

Karat, G.K., 1991: Farmers Survival Strategies in a Drought-prone District: A Village in Chiradurga district, *ADRT, Institute of Social Change*, Bangalore.

Kazi, M.F. and Wodeyar, A.K., 2001: Agriculture in Flood Prone Areas: A Case Study of North Central Region of Bangladesh, *Indian Journal of Regional Science*, Vol. 33, No. 1.

Kulkarni, S.A., 2005: *Looking Beyond Eight Sprinklers*, G.B. Pant University, Pantnagar.

Kumar, A., 2002: Water Supply Scheme in Vidarbha, *Economic and Political Weekly*, Vol. 42.

Kumar, V., and Pal, M., 1985: Development of Irrigation in Haryana: Limitation and Prospects: A Review, *Transaction of Institute of Indian Geographer*, Vol. 7, No. 2.

Mehta, H.M., *et al.*, 1993: *Comparative Study of Sprinkler Vs. Flood Irrigation in Summer Groundnut*, in Central Board of Irrigation and Power (CBIP).

Memon, K.A. 1995: Micro-Irrigation: Profits in the Pipeline, *India Today*, May 1995.

Mohammad, N., 1996: *New Dimensions in Agricultural Geography*, Vol. I-VIII, Concept Publishing Co., New Delhi.

Nagrajan, R., 2003: *Drought: Assessment, Monitoring, Management and Resources Conservation.* Capital Publishing Co., Delhi.

Narayanmoorthy, A. and Desponde, R.S., 2003: Irrigation Development and Agricultural Wages: An Analysis Across States, *EPW*, Vol. 38, No. 35.

Narayanmoorthy, A. 1995: Beneficial Impact of Drip Irrigation: A Study Based on Western India, *Water Resource Journal*.

Narayanmoorthy, A., 2001: *Impact of Drip Irrigation on Sugarcane Cultivation in Maharashtra*, Agro-economic Research Centre, Pune.

Narayanmoorthy, A., 2004: Drip Irrigation in India: Can it Solve Water Crisis, *Development Policy Review*, Vol. 22, No. 4.

Rajgopal, A. 1998: Sprinkler System of Irrigation for Water Efficiency in Scarcity Areas, in Verma, V.V.I. and Rao, A.R.G. (eds.): *Micro-Irrigation and Sprinkler Irrigation Systems*, New Delhi.

Rosegrant *et al.*, 2002: *World Water and Food to 2020*, International Water Management Institute, Colombo.

Rosergrant, W.M., 1997: Water Resources in Twenty-First Century: Challenges and Implication for Action, *International Food Policy Research Institute*, Washington, DC.

Saleth, R.M., 1996: *Water Institutions in India: Economics, Law and Policy*, Common Wealth Publishers, New Delhi.

Seckler *et al.*, 1998: World Water Demand and Supply 1990 to 2025: Scenario and Issue, *Research Report*: 19, International Water Management Institute, Colombo.

Seckler *et al.*, 1999: Water Scarcity in Twenty-First Century, *International journal of Water Resources Development*, Vol. 15, No. 1-2.

Sen, A.K., 2000: *Development as Freedom*, Oxford University Press, New York.

Sengupta, A., 2002: Water Resources and Agricultural Productively in Malda Distinct, *Geographical Review of India*: Vol. 64, No. 3, Kolkata.

Sephardic, S. 1997: Andhra Pradesh: River Water Politics in Drought-prone Telangana, *EPW*, Vol. 32.

Shafi, M., 1990: Agriculture and Rural Transformation in India, *The Geographer,* Vol. 38, No. 1.

Singal, A.M., 1995: "Approaches to Water Resource Development in Arid Areas of Haryana" in Chauhan, T.S. (ed.) *Irrigation in Arid Lands*, Print Well Pub., Raipur.

Singh, A.L., 1992: Impact of Different Source of Irrigation on Cropping Pattern and Farm Practices, *Geographical Review of India*. Vol. 54, No. 1, Calcutta.

Singh, H.P. *et al.*, 2001: *Micro-Irrigation Central Board of Power and Irrigation*, New Delhi.

Sivanappan, R.K., 1987: *Sprinkler Irrigation*, Oxford and IBH Publishing Co., New Delhi.

Suigh, T. 1998: Drought Disaster Management: A Indian Perspective, *National Geographical Journal of India*, Vol. 44, March-December, Nos. 1-4.

Tarjuelo, I.M. *et al.*, 2000: Modelling Evaporation and Drift losses in Semi Arid Region., *Agricultural Water Management*.

Vaidyanathan, *et al.*, 1994: Impact of Irrigation on Productivity of Land, *Journal of Indian School of Political Science,* Vol. 6, No. 4.

Verma, C.V.J. and Rao, A.R.G. (eds.) 1998: *Micro-Irrigation and Sprinkler Irrigation System*, Central Board of Irrigation and Power, New Delhi.

2
Research Design, Data Base and Methodology

2.1 Introduction

The study is based on primary and secondary data. The secondary data have been collected from published and unpublished records of the Government and semi-Government organisations and also from the NGOs and various websites. Some of the important latest publications used for secondary data are:

- India Meteorological Department, Pune.
- Primary Census Abstract of the districts of Haryana, 2001 and 2011.
- General Population Tables of Haryana, 2011.
- Statistical Abstract of Haryana, 2008 to 2011.
- Indian Agricultural Statistics, Vol. I & II, Ministry of Agriculture.
- Agricultural Statistics at a Glance, 2006 to 2010.
- District Statistical Abstract of Hisar, Bhiwani and Mahendragarh, 2007-08.

2.2 Research Design and Data Base

For the collection of primary data multi-stage sampling technique has been adopted. Firstly, on the basis of the physical and socio-economic parameters the drought-prone area has been identified which includes Nangal Choudhary, Ateli Nangal, Narnaul, Kanina, Mahendragarh, Loharu, Badhra, Kairu, Tosham, Siwani, Hisar-II parts of Hisar-I, and Charkhidadri Community Development Blocks. According to the size of the blocks 30 villages in all have been selected from different blocks on the following bases:

(i) It should be basically an inhabited village;
(ii) The villages should be representative of its community development block;
(iii) Some villages should be easily accessible, while others should be inaccessible from remote area;
(iv) Some villages should be close to urban centres, while some should be away from it;
(v) It should have agricultural and allied activities as its major economic activity;
(vi) There should be different size of holdings;
(vii) The inhabitant of the village specially the farmers should be of different socio-economic structure; and
(viii) The village should have sizeable number of households.

After deciding the criteria for selection of villages, 30 tentative locations have been identified and from each location tentatively 4-5 villages have been identified with the help of general maps, available topographic sheets and Administrative Atlas of Haryana (2006). The basic characteristics of the tentatively selected village have been found out from the Primary Causes Abstract of each district. At this stage 2-3 villages with their locations and characteristics have been chosen. Out of these villages, finally, one village has been selected from each Community Development Block after verifying all the details and their characteristics with the administrative authorities and the local people during the course of fieldwork (Fig. 2.1).

Finally, from each of the final selected villages 5 to 10 per cent of the households have been selected on the basis of the socio-economic structure of the household such as age and sex composition, literacy status, size of holdings, status of irrigation, standard of living, irrigation system, caste and religious composition (Table 2.1). In order to compare more precisely the impact of sprinkler irrigation on agricultural development and socio-economic transformation, the non-users of sprinkler irrigation have also been selected from each village. Nearly 29 per cent selected respondents are non-users and 71 per cent are users of sprinkler irrigation. At the end either the head of the household or the person who takes decision in family and agricultural matters have been interviewed according to the well structured questionnaire. A comprehensive questionnaire containing several questions on different relevant aspects, such as, personal and household characteristics of respondents, farming characteristics of respondents, pattern of agricultural landuse, agricultural input and outputs, irrigated area, number of sprinkler sets, irrigated area by sprinklers etc. has been prepared. This questionnaire has been personally along with the assistant have been administered for limited number of respondents in the reconnaissance survey in 3 villages, viz. Mirpur, Ushampur and Bakhra, lying in Hisar-II, Mahendragarh and Bhiwani development blocks respectively.

During the course of reconnaissance survey, the discussion with local influencial and literate peoples and farmers, administrative officers, local authorities, and our own experience and observation in administering the questionnaire, it has been found that the questionnaire needs to be modified and a separate brief questionnaire is required for the general information about the selected village as a whole.

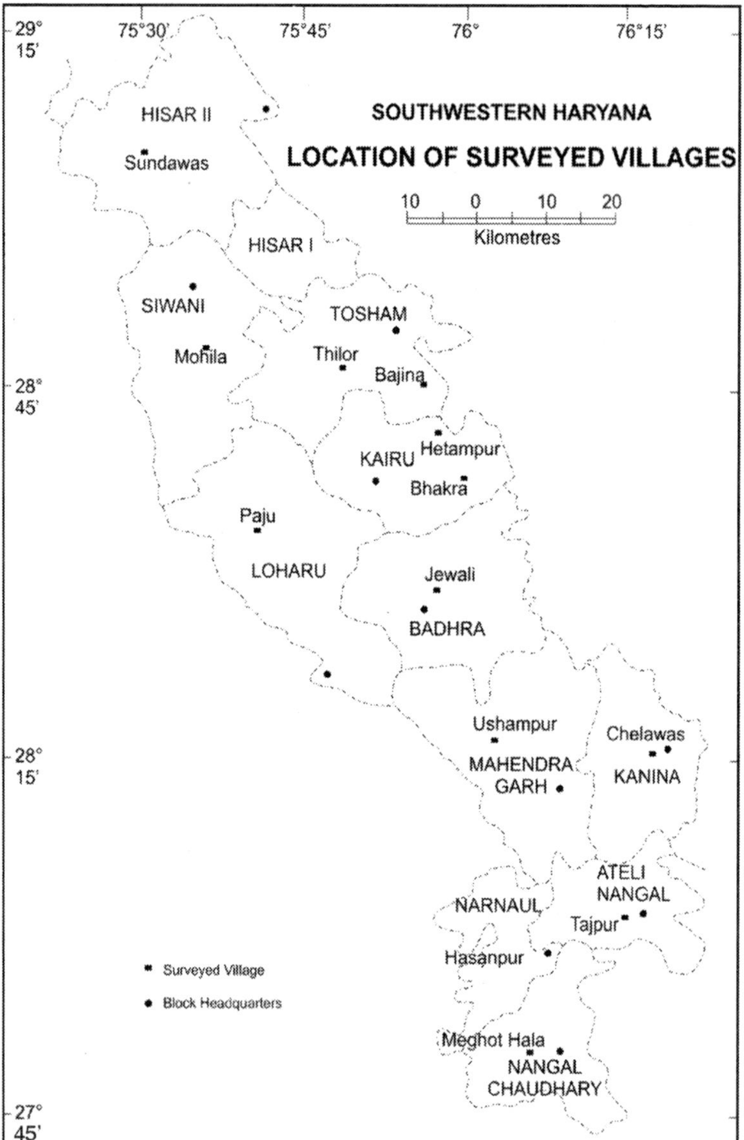

Fig. 2.1: Location of Selected Villages

Table 2.1: Administrative Location of Selected Villages

Sl. No.	*Name of Village*	*Blocks*	*Tehsils*	*Districts*	*No. of Surveyed Households*
1.	Sundawas	Hisar	Hisar	Hisar	40
2.	Mohila	Siwani	Siwani	Bhiwani	30
3.	Thilor	Tosham	Tosham	Bhiwani	35
4.	Bajina	Tosham	Tosham	Bhiwani	50
5.	Hetampur	Kairu	Bhiwani	Bhiwani	45
6.	Bhakra	Kairu	Bhiwani	Bhiwani	25
7.	Paju	Loharu	Loharu	Bhiwani	35
8.	Jewali	Badhra	Dadri	Bhiwani	50
9.	Ushampur	Mahendragarh	Mahendragarh	Mahendragarh	20
10.	Chelawas	Kanina	Mahendragarh	Mahendragarh	40
11.	Tajpur	Ateli Nagal	Narnaul	Mahendragarh	45
12.	Hasanpur	Narnaul	Narnaul	Mahendragarh	25
13.	Meghot Hala	Nagal Chaudhary	Narnaul	Mahendragarh	25

Therefore, a final comprehensive questionnaire containing the following sections has been prepared:

Section-I Personal and household characteristics of respondents including age, sex, education and family size;

Section-II Farming characteristics including size of holdings, patterns of agricultural landuse, adoption of agricultural innovations, agricultural inputs, and yield of crops;

Section-III Sources of irrigation, area irrigated, number of sprinkler sets, type of sprinkler set, area irrigated by sprinkler etc.

Section-IV Social indicators of transformation such as educational status, attitude towards jobs, marriage, family planning, political awareness and involvement, social mobility etc.;

Section-V Economic indicators of transformation such as exposure to mass media income level, standard of living, occupational structure, and housing condition; and

Section-VI Personal experience and observation of the interviewer.

Beside this a separate questionnaire for seeking general information of the village has also been prepared.

2.3 Method of Data Processing

After collecting all the relevant primary and secondary data, the processing of secondary data has been taken-up first. In the first stage, the data have been streamlined and conceptual ambiguities have been removed. Since, the South-western Haryana is a physical unit, its boundaries do not coincide with either district, tehsil or community development blocks. Therefore, the villages falling in the physical unit of Drought-prone Areas of Haryana as identified in chapter 3 have been included in the study region. In this process 2 villages of Dadri, 5 villages of Adampur, many villages lying in south-western portion of Hisar-I community development blocks have been included in the study area. Therefore, the secondary data pertaining to population parameters, landuse, sources of irrigation, number of sprinkler sets, agricultural landuse have been adjusted according to the information, data and records of the respective villages.

Most of the climatic and demographic data found in semi-processed or processed form have been used with little change and calculations. The agricultural data pertaining to general land-use and agricultural landuse have been processed using simple percentage technique. Among various methods of crop combinational analysis, Doi's Critical Table for Crop-Combinational Analysis has been used in this study (Doi, 1959; Mohammad, 1977; Rafiullah,

1965; Weaver, 1959; Ayyer, 1969; Kostrowicki, 1972; and Majeed, 1992).

The intensity of cropping has been worked out using the following formula:

$$\text{Intensity of cropping} = \frac{\text{Gross cropped area}}{\text{New sown area}} \times 100$$

The level of agricultural development has been calculated by standardizing various indicators of development using the following formula of 'Z' Score:

$$Z_{ij} = \frac{x_{ij} - x_i}{\delta x_j}$$

where,

Z_{ij} = Standard value of the indicator;

x_{ij} = Value of x_j variable in the ith observation;

x_i = Mean value of x variable; and

x_{ij} = Standard deviation of x_j variable.

The index of level of social and economic transformation and the composite index of socio-economic transformation have been worked out with the help of the following formula:

$$CIT = \frac{\sum_{x=1}^{n} x_{ij}}{n}$$

where,

CIT = Composite index of transformation;

x_{ij} = score value of ith indicator; and

n = Number of indicators.

The primary data, which are the basic foundation of this research, have been processed and analysed with utmost care and concentration. Firstly, all the entries in the questionnaire have been thoroughly checked and verified. They have been compared with the information of other villages also. The entire questionnaires have been streamlined and discrepancies, if there were any, have been removed in consultation with the contact persons in the villages. If it has not been solved, the village has been revisited to verify the facts or the village has been changed. In this study Mirpur village of Hisar Block has been replaced with Sundawas of the same block.

After streamlining the primary data, most of these have been processed using simple percentage method. The data of different nature and character have been streamlined using scoring scheme. Although, there is always an element of subjectivity in scoring scheme, but all precautions have been taken into consideration to give judicious score to each element.

The literacy level of the respondents and households has been ascertained using the following scoring scheme.

Literacy Standards	*No. of Scores*
Illiterate	0.00
Informal education	0.25
Primary school	0.50
Middle school	0.75
High school	1.25
Senior Secondary school	1.75
Graduation	2.50
M.A. and Others	5.0

Using the above scoring scheme the literacy level of each respondent has been ascertained and they have been classified in different categories. Similarly, on the basis of the same scoring scheme the composite index of education of each household has been ascertained using the following formula (Mohammad, 1988 and Kanika, 2007).

$$CLIH = \frac{\sum_{i=1}^{n} x}{n}$$

where,

CLIH = Composite literacy index of the household;

x = Total educational score of the household; and

n = Total members in the households above 5 years of age.

Based on this composite literacy index of households, they have been classified in different categories.

The processed data have been presented in the form of tables, diagrams, maps, and charts, according to the requirement of the data using computer and GIS software.

References

Ayer, N.P., 1969: "Crop Region of Madhya Pradesh: A Study in Methodology". *Geographical Review of India*, Vol. XXXI, No. 1, p. 17.

Bhatia, S.S., 1965: Patterns of Crop Concentration and Diversification in India, *Economic Geography*, Vol. 43.

Census of India, 2011: *Haryana, Administrative Atlas*, Directorate of Census.

Census of India, 2001: *Haryana-Administrative Atlas*, Directorate of Census.

Doi, K., 1959: The Industrial Structure of Japanese Prefecture, *Proceeding of International Geographical Union,* 1957, Regional Conference in Japan.

Indian Agricultural Statistics, Vols. I & II, Ministry of Agriculture, Government of India, New Delhi.

Kendall, M.G., 1939: The Geographical Distribution of Crop Productivity in England, *Journal of Royal Statistical Society*, Vol. 162.

Kostrowicki, J., 1972: "Some Method and Techniques to Determine Crop and other Landuse Combinations as Used in Polish Landuse Studies", in M. Shafi, M. Anas and M.F. Siddiqui, *Symposium on Landuse in Developing Countries, 21st International Geographical Congress*, Department of Geography, A.M.U., Aligarh.

Mahmood, A., 1977: *Statistical Methods in Geographical Studies,* Rajesh Pub., New Delhi.

Majeed, A., 1992: "Crop Combinational Analysis: A Review of Methodology", in Mohammad, N. (ed.) *New Dimensions in Agriculture Geography,* Volume 5, Concept Publishing Co., New Delhi.

Mohammad, N. and Rai, S.C., 2014: *Agricultural Diversification and Food Security in Uttarakhand,* Concept Publishing Co. Pvt. Ltd., New Delhi.

Mohammad, N., 1988: *Caste and Primary Occupations: A Geographical Analysis,* Concept Publishing Co., New Delhi.

Mohammad, N. and Raza, M., 1992: "Agricultural Meteorological Data Base and Agricultural Productivity" in Mohammad, N. (ed.) *New Dimensions in Agriculture Geography,* Volume (I2), Concept Publishing Co., New Delhi.

Rafiullah, S.M., 1965: "A New Approach to Functional Classification of Towns", *The Geographer,* Vol. XII, pp. 40-53.

Ramachandran, R., 1992: Correlates of Innovations Among Farmers in the Ganga Plain; in Mohammad, N. (ed.) *Human Perception on Technological Change in Agriculture,* Concept Publishing Co., New Delhi.

Shah, S.L., 1985: "Some Methodological and Pragmatic Issues in Land Capability Classification in Himalaya", in J. Singh (ed.), *Environmental Regeneration in the Himalaya: Concept and Strategies,* Gyanodaya Prakashan, Nainital.

Singh, J., 1977: A New Technique for Measuring Agricultural Efficiency in Haryana, *The Geographer*, Vol. XIX (1), Aligarh.

Verma, K., 2007: *Role of Urban Agricultural in Life Support System: A Case Study of Alipur Block, Delhi.* Unpublished Dissertation. Department of Geography, Delhi University, Delhi.

Weaver, J.C., 1954: "Crop-Combination Regions in the Middle West", *Geographical Review,* Vol. XLIV, No. 2, p. 176.

3

Drought Prone Areas of South-western Haryana

Identification and Delimitation

3.1 Introduction

The western part of Haryana covering nearly 50 per cent area of the state experiences semi-arid to arid climatic conditions which results into occurrence of various kinds of drought almost every year. Hence, there is a need to identify and delimit the drought-prone area of the state for rational and scientific planning for its development.

Droughts are one of the most serious natural hazards to mankind. They have been the root causes of famines in many parts of the world and more so in our country where agricultural success and prosperity are vitally dependent on the amount and distribution of rainfall. Droughts are usually defined as periods of dryness due to lack of rain, though the concept varies from place to place depending upon the normal climatic conditions, available water resources, landuse, agricultural practices and various other economic activities in the region.

Drought is a situation of water scarcity. It is generally considered as a recurrent environmental phenomena caused by failure of rainfall and depletion of surface and ground-water.

Wherever, there is a departure of rainfall from normal, drought might occur, but its intensity and severity is maximum in semi-arid and arid regions. Therefore, a large number of people in the world especially in arid and semi-arid areas are affected by drought because of high variability of rainfall and high evapotranspiration. Annually, nearly 19 per cent geographical area and 16 per cent population of our country are vulnerable to drought disaster because a large number of people and their livelihood are dependent on monsoon (Singh, 1988).

3.2 Concept of Drought

The concept of drought is complex, complicated and multi-dimensional. Different scholars have defined drought differently. Meteorologists define drought as the failure of precipitation; agriculture scientists observe drought as famine and people's inability to purchase required entitlements. Hence, there is no consensus over the definition of drought, which could frame our broad concept of drought. So, it is worth understanding the different definitions of drought given by different scholars and institutions. Some definitions are given here:

Warrick (1957) defines it as "a conditions of moisture deficit, sufficient to have an adverse impact on vegetation, animal and man over sizeable area." "Drought is a period of abnormally dry weather sufficiently prolonged for the lack of precipitation to cause a serious hydrological imbalance and carries connotations of a moisture deficiency with respect to man's usage of water (McMohan and Diaz Arenas, 1982). Drought is a condition whenever, the amount of water which has been expected and relied upon for use in any of man's activities cannot be met for some reason. It is a natural reoccurring feature of climate that originates from a lack of precipitation over an extended period of time, usually one season or more. Drought can occur virtually in all climates. The precise definition depends on the region, but the definition is often determined by comparing recent precipitation to a 30-year average. In some areas, precipitation that is only 75 per cent of a 30-year average is considered as drought" (Takeuchi, 1974).

The aforesaid definitions do not give due emphasis on socio-economic dimensions of drought. Periodicity, variability and spatial distribution of rainfall as the causes of drought have drawn much attention, but socio-economic behaviours and responses of people in drought-prone area have been ignored in determining and defining the severity of drought.

When socio-economic, physical and environmental factors are considered together some common characteristics of drought are:

- It is creeping phenomena, unlike floods, it is not a distinct event.
- It is often the result of many accompanying factors.
- It is caused by natural as well as human factors.
- The intensity of impacts and consequences differ from region to region.

So, drought is the environmental as well as socio-economic phenomena which are generally dealt in the time space context. Socio-economic dimension of drought is very important to understand. Socio-economic dimension of drought must incorporate the economic activities and social organization and behaviour of the people in the region.

Drought is a transitory disaster generally caused by prolonged failure of rainfall and water supply. It may also result by prevailing water use habits and water storage practices. It affects adversely the agriculture and other economic activities of the area, leading to production and employment shortfall, drinking water shortage and geographical dislocation of people.

3.3 Classification of Droughts

Droughts have been classified by different scholars in various ways, but a clear understanding and better appreciation of the problem of drought and aridity become possible after the theme of climatic classification came to be developed at the turn of the century. Among many systems and schemes evolved for the purpose, Thornthwaite scheme (1948) has gained worldwide acceptance and recognition.

Thornthwaite (1947) has recognized three kinds of drought on the basis of climatic parameters, i.e. (i) the permanent droughts of the driest climates; (ii) the seasonal droughts in climates having well defined rainy and dry reasons; and (iii) the contingent droughts resulting from irregular and variable occurrence of rainfall.

The National Commission on Agriculture (1976) has classified droughts into following three categories:

(i) ***Metrological drought:*** It is defined usually on the basis of significant decrease in rainfall from the normal average. This value is more than 25 per cent of departure from normal.

(ii) ***Hydrological drought*:** They are caused by prolonged meteorological conditions with a significant depletion of surface water and consequent drying up of reservoir, lakes, streams and rivers and also fall in groundwater levels.

(iii) ***Agricultural drought*:** When soil moisture and rainfall are inadequate during the growing season to support a healthy crop growing to maturity and cause crop stress and wilting, it is called agricultural drought.

On the basis of aridity index, Reddy (1979) has identified four types of drought, i.e. Moderate droughts, Large droughts, Severe drought, and Disastrous droughts. Reddy has further attempted to classify droughts on the basis of physio-socio-economic parameters. After discussing the crucial role of all these parameters on agriculture, he has assessed and correlated these factors for identification and delimitation of drought-prone areas of India. Thus, he has broadly followed three criteria for identification and delimitation of drought-prone areas of India, i.e. (i) climatic characteristics, (ii) soil factors, and (iii) agronomical and socio-economic conditions.

After discussing in depth the role of climatic parameters, soil conditions, potential evapotranspiration, degree of water need, irrigation facilities, arable land, agricultural intensity, crop dominancy, etc. he has identified following three zones of droughts/ drought-prone areas covering 155 million hectare area accounting to 47 per cent of total area of India on the basis of correlation and superimposition of the maps depicting various parameters (Reddy, 1979).

(i) Zone-I: Extreme Drought-Prone Areas
(ii) Zone-II: Sever Drought-Prone Areas
(iii) Zone-III: Moderate Drought-Prone Areas

For identifying the drought-prone areas of south-western Haryana, moisture index, aridity index, climatic parameters, soil factors, agronomical and socio-economic conditions have been taken into consideration. They have been mapped using isopleths and choropleth techniques of mapping. These maps have been superimposed and finally, the drought-prone areas of south-western Haryana have been identified and delineated. It covers the substantial western part of Hisar-II, south-western part of Hisar-I,

Siwani, Tosham, Loharu, Kairu, Badhra, Mahendragarh, Kanina, Ateli Nangal, Narnaul and Nangal Choudhary Community Development Blocks (Fig. 1.1).

References

Mohammad, N. and Rai, S.C., 2014: *Agricultural Diversification and Food Security in the Mountain Ecosystem: A Case Study of Uttarakhand.* Concept Publishing Co. (P) Ltd., New Delhi.

Reddy, N.B.K (ed.)., 1979: *Drought-prone Areas of India*, Rayalseema Geographical Society, Tirupati.

Reddy, N.B.K. and Singh, V.R. 1979: "Delimitation of Drought-prone Areas of India: A Geographical Approach", in N.B.K. Reddy (ed.) *Drought-prone Areas of India*, Rayalseema Geographical Society, Tirupati.

Sen, A.K., 2000: *Development as Freedom*, Oxford University Press, New Delhi.

Singh, T., 1988: Drought Disaster Management: An Indian Perspective, *National Geographical Journal of India*, Vol. 44, No. 1-4.

Subrahmanyan, V.P. and Hema Malini, B., 1979: "Delimitation of Drought-prone Areas: Techniques and Methodology", in N.B.K. Reddy (ed.) *Drought-prone Areas of India*, Rayalseema Geographical Society, Tirupati.

Thornthwaite, C.W. 1947: Climates on Moisture Conservation, *Annals of the Association of American Geographers*, Vol. 37, No. 2.

Thornthwaite, C.W., 1948: An Approach Towards a Rational Classification of Climate, *Geographical Review*, Vol. 38, No. 1.

4
The Geographical Profile of the Study Area

4.1 Introduction

No other branch of human geography is so directly associated with physical environment as agricultural geography. Nature with its diverse conditions offers different possibilities which man opt with his given level of technology (Visher, 1932; Hettner, 1947; McHarg, 1966 and Gregor, 1970). Therefore, a brief account of physical environment of agricultural relevance of South-western Haryana is given here.

South-western Haryana extending over 27° 42′ to 29° 15′ N. Lat. 75° 15′ 10″, to 76° 30′ 15″ E. Long. is a part of the Indo-Ganga plain which lies between Peninsular India in the south and recently built Himalayan mountain in the north. It belongs to the recent period of Geological history and is formed by the two tectonic movements which created a sag between the northern edge of Gondwana land and the recently built Himalayan chain. This shallow depression is filled by the sediments brought by the Himalayan rivers in the north and northern Gondwana land of Peninsular India in the south (Wadia, 1919). The Geodetic Survey of India indicates that the bottom floor of this trough is not an even surface, but is marked with many irregularities and buried ridges. One such ridge runs under south-western Haryana from Delhi following a north-west direction towards the Salt Range of Pakistan. (Aithel *et al.,* 1964 and Wadia, 1919).

4.2 Geomorphological Divisions

On the basis of geological formation South-western Haryana is divided into two main divisions, viz. Indo-Ganga Plain and Aravali System. The Indo-Ganga plain is further divided into Bhangar land and sandy undulating plain. As such South-western Haryana is geomorphologically divided into following geomorphic divisions (Fig 4.1):

4.2.1 Indo-Ganga Plain:
    4.2.1.1 Bhangar Land, and
    4.2.1.2 Sandy Undulating Plain.
4.2.2 Aravali Ridge and Out-Crops.

4.2.1 Indo-Ganga Plain

The Indo-Ganga Plain is formed by the deposition of the alluvium brought by the rivers of the Himalayas in the north and the northern peninsular rivers in the south. It ranges in age from the Pleistocene to the recent times. These are of varying thickness and consist of silt and clay with occasional gravel beds specially in older deposits (Shrivastava *et al.*, 2006). There occurs bed of varying thickness of impure calcareous making interplaying pans of irregular concretion popularly known as *Kankar*. The alluvial plain is rich in salt-content.

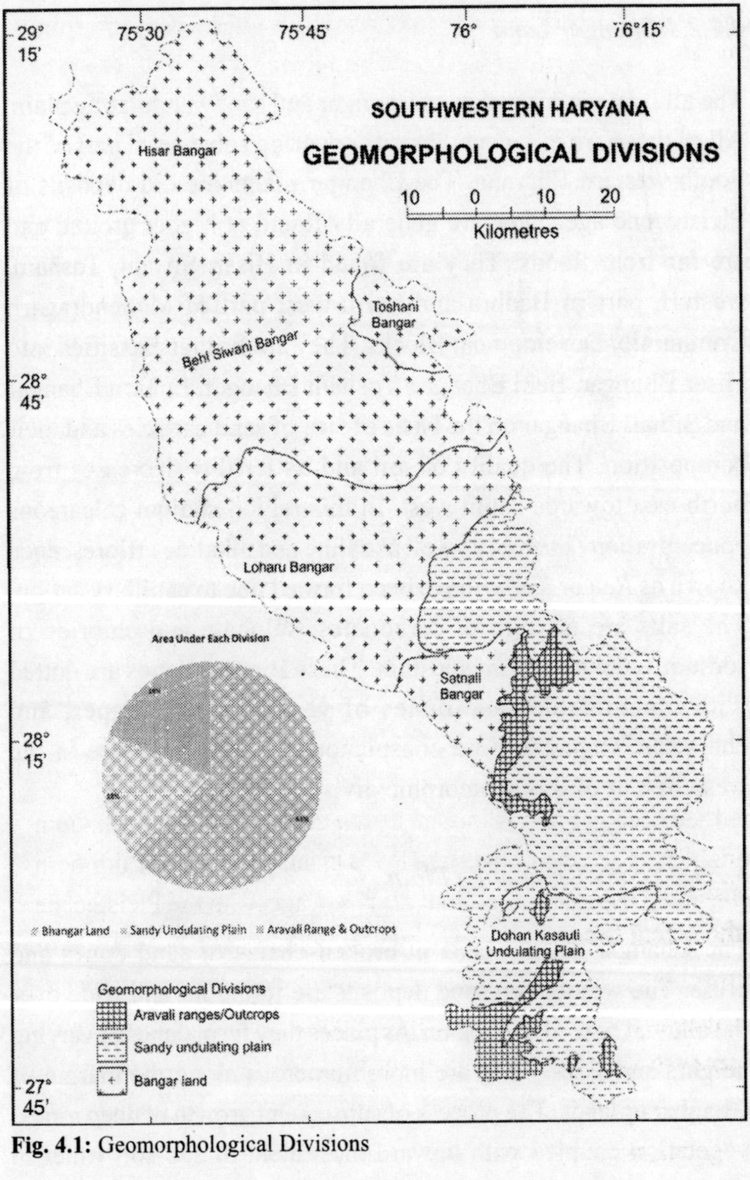

Fig. 4.1: Geomorphological Divisions

4.2.2.1 Bhangar Land

The alluvial plain consists of Bhangar and sandy undulating plain. All of these are of recent deposits covering substantial part of the South-western Haryana. The Bhangar plains are old deposits of Pleistocene age. They are generally found at higher ground and are far from floods. They are found in Hisar, Siwani, Tosham, western part of Badhra and north-west part of Mahendragarh Community Development Blocks. They are further classified into Hisar Bhangar, Behl Bhangar, Tosham Bhangar, Loharu Bhangar and Sitnall Bhangar on the basis of size of sand particles and their composition. The quality of soil and its fertility decreases from north-east towards south-west. In this region sodium calcareous concentration *(kankar)* as well as saline and alkaline efflorescence known as *Reh* or *Kallar* have been formed due to capillary action. The salts are mixture of carbonates, sulphate and chloride of sodium, calcium and magnesium. These Bhangar lands are dotted with innumerable sand dunes of various sizes, shapes, and character. They are more conspicuous by their presence in the western part of this geomorphic division.

4.2.1.2 Sandy Undulating Plain

The Bhangar usually ends in broken chains of sand dunes and cliffs. The windblown sand deposits are found far and wide over the alluvial base of this region. At places they form dunes of varying heights and sizes. They are more numerous along the margin of Ghaghar uplands. The process of subsequent growth of deep rooted vegetation coupled with upward movement of sub-soil water in the loose and absorbent sand material has bounded them together and made them stable (Duggal, 1975).

The sandy undulating plains are found in almost all the

community development blocks of Bhiwani district leaving only the areas of Aravali out-crops. This plain has a windborne topography in which the sand dunes and rocky out-crops stand out distinctly against the level horizon of sand swelling waves. The region is a treeless undulating desert locally known as Bhangar (Krishnan, 1960).

4.2.2 Aravali Ridges and Out-crops

In the southern part of the region specially in Mahendragarh district, geological events of the Aravali cynclinorium are exposed. The lithological units are mainly of Pre-Cambrian quartzite, schist, mica, hyalites etc. They run from south-east to north-west direction with a series of outcrops of different sizes. They are found in Mahendragarh, Nangal Chaudhary and Badhra community development blocks. These are characterised with structural hills and shallow pediments dominating in Mahendragarh district. The Aravali hills are found in a very narrow strips, standing out distinctly against the otherwise level horizon above the sandy plain. They also form a series of flat topped ridges half buried either in the alluvium or other aeolian deposits. These are surrounded by shallow pediments (Singh, 1976).

4.3 Physiographic Division

Physiography exercises direct influence on landuse specially through elevation, ruggedness, slope, slope aspect and soil development. South-western Haryana has a complex physiography and could be divided into following 4 physiographic divisions based on the information obtained from Remote Sensing (Fig. 4.2).

4.3.1 The Alluvial Plain.

4.3.2 The Sand Dunes.
4.3.3 The Undulating Sand Plain.
4.3.4 The Rocky Hills.

4.3.1 The Alluvial Plain

The alluvial plains cover the largest portion of the state. A small portion of it lies in a narrow strip in south-western Haryana. It lies above 300 metres above mean sea level. It is found in the extreme north-eastern part of region covering western parts of Hisar-I and Hisar-II Community Development Blocks. The plain has a gentle slope from north-east to south-west. The monotony of the gentle level plain is disturbed by the occurrence of sand dunes of different types, shape and nature. The inter-dunal plains are generally very fertile having fine particles of silt and sand.

4.3.2 The Sand Dunes

The sand dunes of various shape and size form a thirsty land clothed by steppe vegetation in the western part of the region covering western part of Hisar-II, southern Hisar-I, Siwani, substantial part of Tosham, Kairu, Badhra, a small portion of north-western Mahendragarh and whole of Loharu community development blocks. The overall elevation in less than 300 metre above the level of the ground. Some of the dunes are moving, while most of them are stationary. Longitudinal ridges like dunes are common with interspersed crescent dunes. The rocky projections of Aravali disturb the monotony of this region. The mobile sand dunes at places present the serious problem of erosion by wind and threaten to impair the posterity of fertile alluvial plain lying to their north and north-east. This area is going under change due to increasing irrigation facilities specially the sprinkler irrigation which is

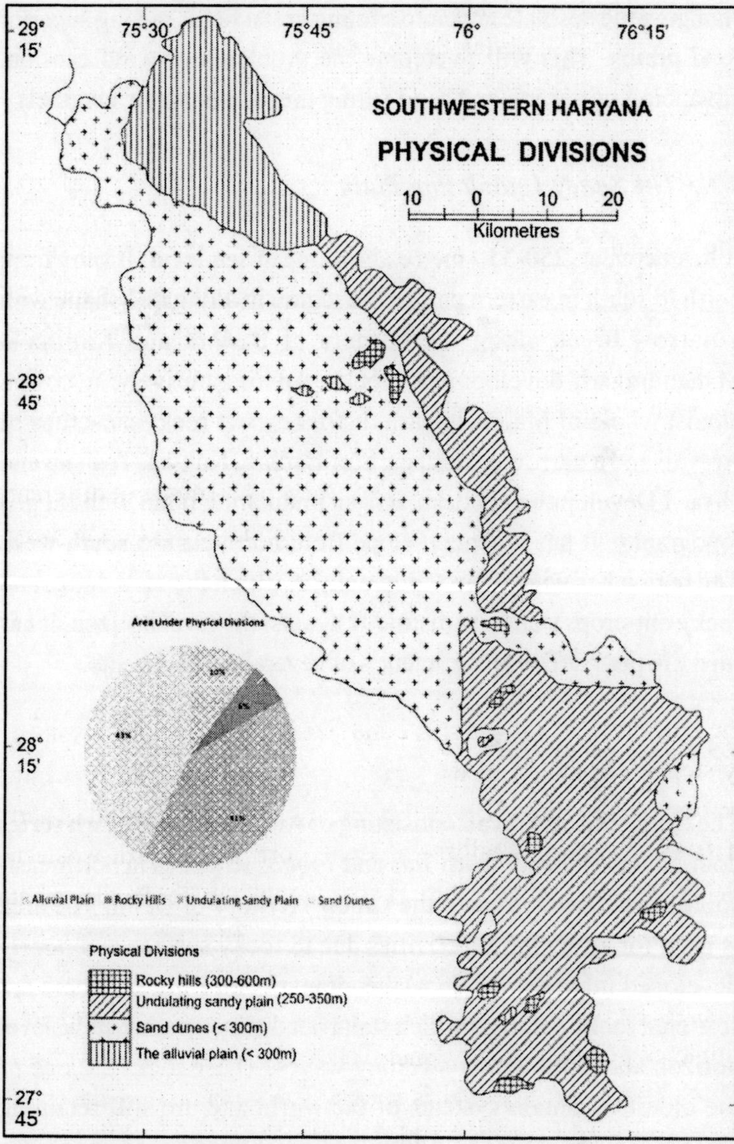

Fig. 4.2: Physical Divisions

changing the desolate area into smiling green fields raising superior food grains. This will overcome the problems of wind erosion, advancing of deserts and sand filling menace of water channels.

4.3.3 The Sandy Undulating Plain

It lies between 250-350 metre above mean sea level. It runs from north to south in eastern part of the region in alongated shape with a narrow break along the borders of Badhra and Kanina—Mahendragarh development blocks. Administratively, it covers almost whole of Mahendragarh district expect rocky out-crops of Aravali and a narrow eastern parts of Badhra, Kanina, Tosham and Hisar-I Development Blocks. It is an undulating plain with rugged topography. It has a general slope from north-east to south-west. The region has wind worn topography in which the sand dunes and rocky out-crops stand out distinctly against the level horizon. It has infertile poor soil with deficiency of several soil nutrients.

4.3.4 The Rocky Hills

The extension of Aravali consisting of Alwar and Ajaibgarh series dominating Mahendragarh *Tals* and *Tabbas* stretches in north-east-south-west direction up to the Union Territory of Delhi. The area is unfit for agricultural activities due to rocky surface and poorly developed infertile soil. In Mahendragarh and Bhiwani, there are scattered rocky outcrops which stand out distinctly against the level horizon above the sandy swelling waves. These rocks are one of the oldest mountain systems of the world and are still retaining some relief. These bare bold rounded formations present a curiously wind worn topography which illustrates the phenomena of desert erosion. On the whole, they form series of flat topped ridges, gaunt, bony figures half buried in the alluvium or aeolian deposits. Thus, it represents a local relief. All these hills are dissected by generally dry, but at a time torrential filled *nalas*. The water table is low and there is almost absence of good soil cover.

4.4 The Climate

The Climatic parameters along with adaphic condition play a determining role in shaping the patterns of landuse/land cover. Climate provides possibilities of raising variety of crops and affects their yield in conjunction with adaphic and socio-techno-economic conditions. Therefore, there is a need to give a brief relevant account of the climatic conditions of south-western Haryana.

The Climate of south-western Haryana is arid to semi-arid and is characterised by a lot of variations in temperature, rainfall and relative humidity. The variable nature of climatic parameters has a strong bearing on pattern of landuse specially of agricultural landuse and cropping system. These conditions prevail mainly due to its continental location, distance from mountain and sea. However, the climate of south-western Haryana may be summarized as a sub-tropical continental monsoon type having the characteristics of seasonal rhythm, hot summer, cool winter, dry conditions, meagre, and unreliable absent rainfall and great annual range of temperature. There are a lot of variations in mean minimum, mean maximum and average annual temperature during different seasons of the year. Although mean minimum annual temperature represents a criss-cross and complicated pattern, but mean maximum and annual average temperature show the clear cut pattern. The mean maximum annual temperature decreases from south-west to north-east. Hisar, Bhiwani and Mahendragarh experience more than 25°C of mean annual temperature. Narnual and Loharu record more than 32°C average temperature, while it is less than 32°C in Bhiwani and Hisar. The similar pattern in

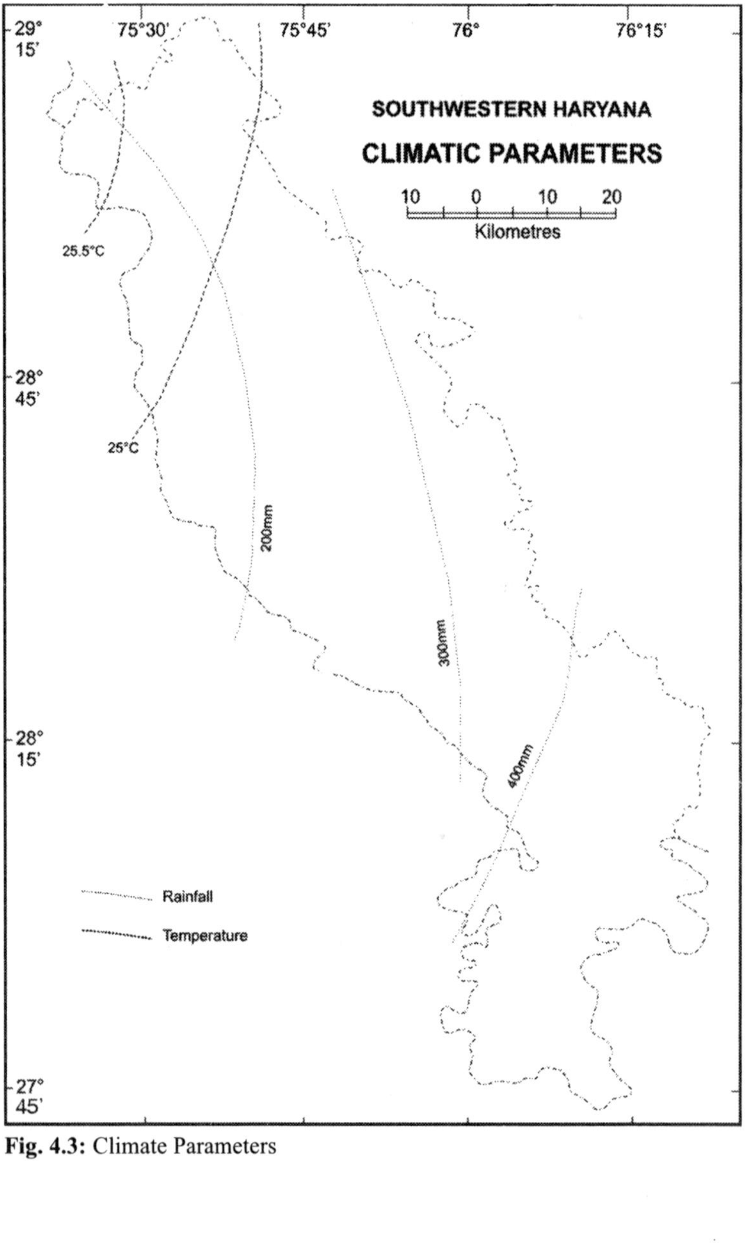

Fig. 4.3: Climate Parameters

noticed in case of average temperature, but whole of south-western Haryana experiences more than 25°C temperature which decrease towards north-east. This distributional pattern is mainly due to its proximity with Thar desert of Rajasthan in the north-east.

The distribution of rainfall is highly irregular both in space and time and water deficiency is a problem. There is a marked difference in the distributional pattern of annual rainfall due to prevalence of easterly moving monsoon and westerly moving disturbances. The average annual rainfall is less than 300 mm. in north-western Haryana specially in Loharu and Siwani, while in the eastern part it is more than 300 mm. such as in Hisar, Bhiwani and Mahendragarh.

There are a lot of seasonal variation in occurrence of rainfall in South-western Haryana. Nearly, 80 per cent rainfall is received from western disturbances during the months of January and March. The monsoonal rainfall decrease from north-east towards south-west, while opposite is the case with winter rainfall. This is very beneficial for *rabi* crops. The monsoon rain sets the agricultural season starting from *kharif*. The intensity of rainfall is more in western part of the region as compared to eastern region. It is more in monsoon season than cyclonic period. Most of the rainfall is wasted in runoff, while the cyclonic winter rainfall generally falls in light drizzle which allows sufficient time for its absorption in the soil. There is considerable variation in the monthly and working regime of rainfall from place to place depending upon the distance from the mountain, topographical alignment and location with reference to desert. However, the coefficient of variability of annual rainfall is 30 to 50 and its distributional pattern is just opposite to the distributional pattern of rainfall, i.e. higher the rainfall, lower is its variability. It is highest in south-western part and it decreases towards north-east. The average annual potential evaporation is around 143 cm which decreases towards east, while the average annual water deficit is around 143 cm which also decrease from south-west towards north-east.

4.5 Soils

South-western Haryana has a variety of soils having a complex mosaic distributional pattern which is reflected at every level. It is mainly due to a variety of parent materials, variable amount of rainfall, intensity of temperature and local as well as regional pattern of soil formation. Variation in the anthropogenic determinants such as intensity of irrigation, landuse, hydrological character, etc., have also a strong bearing on the pattern of soil distribution in the region. Traditionally, soils used to be classified into *Urvarak* (fertile) and *Anurvarak* (infertile) soils in ancient India. It has been gradually modified on scientific lines, and soils have been classified on the basis of rainfall, and chemical and physical characteristics of soil. The south-western Haryana falls into the zone of Acidic and Torrid soil groups.

Earlier classification of soil were also based on water required process of soil formation and their properties, but in the present study, more comprehensive classification suitable for planning agriculture has been done based on *p*H value, electric conductivity or presence of soluble exchangeable salt, soil texture and such other characteristics. It will help in assessing the potentiality for crop productivity and sustainability. Therefore, on these bases the soils of Haryana have been classified into 12 classes grouped into six categories. In south-western Haryana where the Great Indian Desert makes an entry, severe aridity prevails. The most pre-dominant component in these stretches is desert sand of quartzite origin having well rounded grain with a fair proportion of calcium. However, the soils of south-western Haryana fall into three main categories with five following sub-classes based on their nomenclature. The spatial pattern of distribution of these soils is shown in Fig. 4.4:

4.5.1. Coarse Sandy Soil
4.5.2. Loamy Sand Soil
4.5.3. Sandy Loam Soil
4.5.4. Fine Loamy Soil
4.5.5. Rocky Surface and Shallow Piedmonts

4.5.1 Coarse Sandy Soil

The Coarse sandy soils occur in northern part covering parts of Hisar-I, and Hisar-II, whole of Tosham, Kairu, and two-thirds northern part of Loharu Community Development Blocks. They consist of large size of sand particles, and are dotted with sand dunes. The pattern of dune frequency varies over the dune field, and inter dune soils are mostly loamy sand. The area is subjected to excessive wind erosion to overlaying by drifting sand cover aeolian sand. These are very loose soils with maximum permeability causing excessive drainage, a tendency to dry-up rapidly and low water holding capacity. They are deficient in soil moisture. They are suitable for coarse grains like *jowar*, *bajra*, barley, maize etc. But, if proper irrigation water with adequate amount is managed, it can be very productive.

4.5.2 Loamy Sand Soil

The Loamy sand soil is found in the south of coarse sandy soil. It covers the southern part of Mahendragarh Community Development Block. Here, the soil particles are small and structure and texture of the soil is fairly good. It is interrupted with variety of dunes of various size and shapes. The soils have single grained

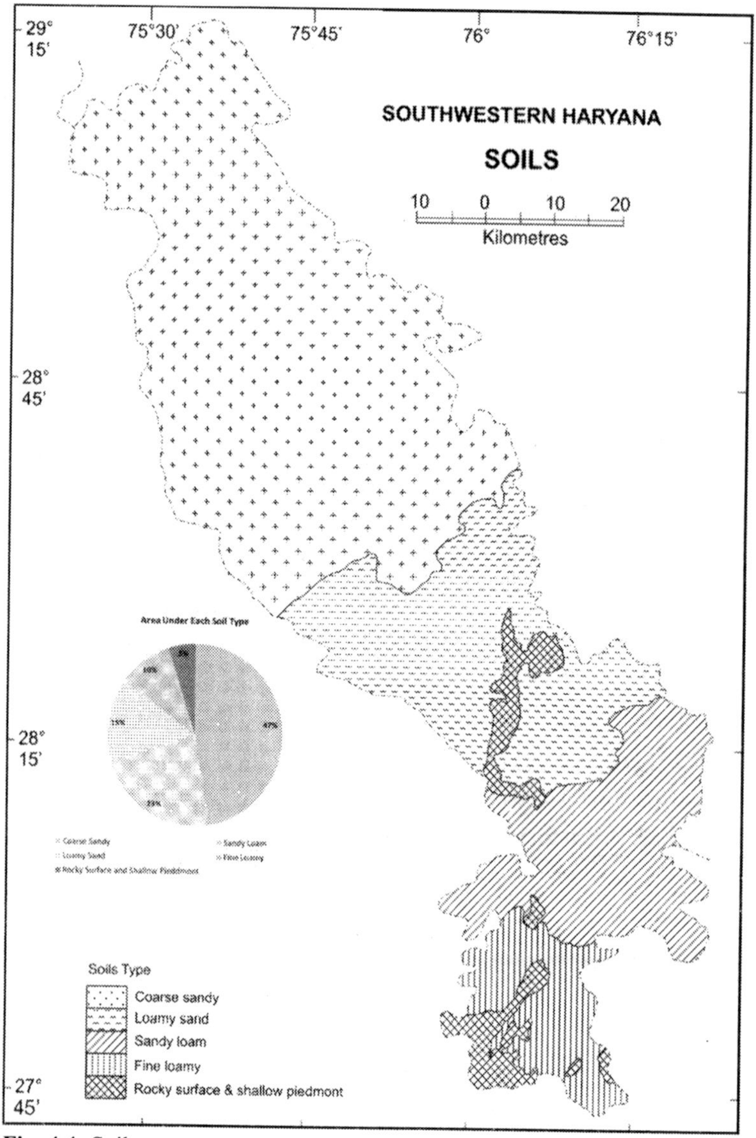
29° 15'
75°30'
75°45'
76°
76°15'
SOUTHWESTERN HARYANA
SOILS
10 0 10 20
Kilometres
28° 45'
28° 15'
27° 45'
Area Under Each Soil Type
Soils Type
Coarse sandy
Loamy sand
Sandy loam
Fine loamy
Rocky surface & shallow piedmont

Fig. 4.4: Soils

structure, because the individual particles exist independently and are predominantly of fine texture. The interdunal plains are very conspicuous by their presence. The dune uplands where rainfall meagerness and truancy are generally inimical to cultivation are characterized by pedocol seirozem. They are also very porous soil which allow water to flow through them almost unimpeded. These feature cause crops to suffer in periods of drought. The deficiency of moisture is main cause of poor agriculture produce, but if it is supported with adequate irrigation water, it will give not only high yield, but will be suitable to fine food grains crops.

4.5.3 Sandy Loam Soil

Sandy loamy soil is found is Kanina, Ateli Nangal, southern portion of Mahendragarh and northern part of Narnaul community Development Blocks. It consists of larger size of sand particles and gravels of different sizes. It is poor in humus and organic matter. It has thin layer of soil with large number of exposed rocks. It is normally devoted to the cultivation of coarse grain crops like *jowar*, *bajra*, maize etc. In the recent times in the eastern part of this region, small area is devoted to cultivation of rice, if irrigation facilities are available. The western part is more than barren and rocky and is not suitable for agricultural activities.

4.5.4 Fine Loamy Soil

The fine loamy soil is found in the extreme southern part of the region in the Narnaul and Nangal Chaudhary Development Blocks. It is soft soil with almost ideal proportion of sand, silt and clay. In the district of Mahendragarh, it resembles more with the inferior loamy soil locally known as *bur,* but in the remaining areas it is similar to light loan soil. It has an ideal degree of both permeability and water holding capacity which can further be improved by the use of farmyard manure or green manuring with cluster been or *dhaincha*. The soil gives good response to irrigation, though, like sandy soils, it needs the help of modern practice to build-up. With sprinkler irrigation and a moderate amount of rainfall, the seed germination is rapid and crops yields are high.

4.5.5 Rocky Surface and Shallow Piedmonts

The rocky surfaces and shallow piedmonts are in fact the continuation of Aravali. The exposed surface is highly concentrated in Mahendragarh and Nangal Chaudhary community development blocks. Besides these, few patches are also found in Narnaul and Ateli Nagal community development blocks. These soils are coarse and are not sufficiently thick and favourable for crop production. It is gradually vanishing as it is getting washed away by torrential summer rains and is removed off by wind blast of hot dry periods. It is too shallow, stony, infertile to be used for raising of field crops. It is a residual, weathered material lying over present rocks and is very hungry soil. Under it, there is a collar of often deep and coarse light inferior soil with below average fertility and low base status. As such these are very infertile poor soils which need rigorous conservation and management practice to make it suitable for crop production.

The soils of south-western Haryana suffer with the major problems, viz. salinity and alkalinity, waterlogging, soil erosion, and deficiency in chemical and biological nutrients. They have decidedly depressing effect on crop yield.

4.6 Socio-Cultural Setting

South-western Haryana covering an area of about 6429 km^2

supports a population of 21,62,222 with an average density of 336 person per km^2. The population is unevenly distributed and the density of population decrease from north-east to south-west. The density of population, i.e. 428 person per km^2 has been recorded in Narnaul and Mahendragarh, while the lowest, i.e. 125 person per km^2 is recorded in Siwani community development block. It is a male dominated region with 11,46,752 (53%) male and 10,15,470 (47%) female which is unevenly distributed. It is inhabited by predominantly rural population which accounts nearly 80 per cent of the total population leaving only one-fifth as urban population. The literary level is considerably high with nearly 67 per cent as literate population which varies from 61.7 per cent (lowest) in Siwani to 70.57 per cent (highest) in Mahendragarh block.

4.7 Economic Setting

The region has basically an agrarian economy with nearly 70 per cent population engaged in agriculture directly or indirectly. Nearly 82 per cent of the total area is under cultivation, while nearly 87 per cent is culturable (Table 4.1). The total cropped area amounts to 9,72,500 ha giving 188 per cent intensity of cropping. The per cent of net sown area decrease from north-east to south-west. The forest cover is almost negligible covering only 0.71 per cent of the total area. Nearly 12.5 per cent is not available for cultivation which includes area under settlements, roads, canals etc., on the one hand and mountain rocky barren land on the other. The former is highly concentrated in eastern parts of the region, while the latter is mostly found in Mahendragarh and Narnaul community development blocks. Like forest, the other uncultivated land excluding fallow land covers a negligible percentage of the area.

Table 4.1: Spatial Patterns of Landuse/Land Cover (2005-06)

Categories	*Area in ha*	*% age to*
Forest	4,442	0.71
Land not available for cultivation	78,400	12.49
Other uncultivated land excluding fallow	1,400	0.22
Fallow land	27,225	4.34
Net sown area	5,16,133	82.24
Total	**6,27,600**	**100.00**

Note: The data have been collected, adjusted and processed on the basis of village records and information collected by the author.

There are two main agricultural seasons in the region, viz. *kharif* season—the season of summer crops, and *rabi* season—the season of winter crops. Besides these, there is a zaid season—the season of additional crops. In this season very few crops are grown. The major *kharif* crops are *jowar*, *bajra*, rice, and cotton, while wheat, barley, gram and oilseeds are the main crops of *rabi* season (Table 4.2). Cereals alone occupy 4900 ha amounting to 50 per cent of gross cropped area, while pulses cover 9,1400 ha or 18.65 per cent of gross cropped area.

Table 4.2: Spatial Cropping Pattern (2005-06)

Crops	*Area in ha*	*%age to*	
		NSA	GCL
Rice	8,300	1.61	0.85
Bajra	2,98,800	57.89	30.72
Wheat	1,50,100	29.08	15.43
Barley	7,000	1.36	0.72
Gram	90,200	17.48	9.28
Other Pulses	91,200	24.17	9.38
Oil seeds	1,24,800	13.87	12.84
Cotton	71,600	7.36	7.36
Others	1,30,500	25.28	13.42
Total	**9,72,500**	—	**100.00**

Note: Data have been collected, adjusted and processed on the basis of village records and information collected by the author.

The areal coverage of oilseeds is more in north-western part of the region. *Bajra* is the major crop of the region. It occupies 2,98,800 hectare amounting to nearly 30.72 per cent of gross cropped area. It is widely cultivated almost all over the region. Wheat is the second major crops of the area. It cover 1,50,100 ha or 15.43 per cent of gross cropped area. It is also widely grown with its more concentration is south-western part of the region. Gram, being third major crop covers 90,200 ha amounting to about 9.38 per cent of gross cropped area. The cultivation of pulses and oilseeds is gaining popularity in the region due to their shortage at national level, and consequently launching of many schemes for their development. Barley, cotton and sugarcane are the other crops grown in the region, but they do not cover cognisable area. The region is not very rich in mineral base, but the industrial base is comparatively strong.

References

Abraham, T. and Lamathi, E. 'Eco-organic Farming' in Pruhit, S.S. and Gehlot, D.C. (eds.): *Trends in Organic Farming in India*, Jodhpur.

Alagh, Y.K. (ed.) 2003: *Globalisation and Agricultural Crisis in India*, Deep & Deep Pub., New Delhi.

Auden, J.B. and Roy, P.C. 1942: Report on the Sodium Salt in Reh in U.P., *Records* of *Geological Survey of India*, Calcutta.

Blanford, H.F., 1889: *Climate and Weather of India, Ceylon and Burma*, London.

Directorate of Economics and Statistics, Ministry of Agriculture, *Agricultural Statistics at a Glance,* 2005 to 2012.

Duggal, S.L., 1975: *Soil Geography of Haryana,* Hisar Agricultural University, Hisar.

Gregory, H.F., 1970: *Geography of Agriculture: Themes in Research*, Prentice Hall.

Hetner, A. 1947: *General Geography of the Humans*, Vol.1, Kohlhammer.

Kendrew, W.A., 1961: *The Climates of the Continents*, Oxford University Press, London.

Klages, K.H.W., 1958: *Ecological Crop Geography*, McMillan Co., New York.

Krishnan, M.S. 1960: *Geology of India and Burma,* Madras.

McHarg, 1966: 'Ecological Determinism' in Darley, F.F. and Melton, J. (eds.) *Future Environment of North American*, New York.

Ministery of Environment and Forest, 2010: *National Wetland Atlas of Haryana*, Ahmedabad.

Pascoe, E.H., 1973: *A Manual of Geology of India and Burma-III,* Government of India, Delhi.

Pretty, J., 2006: *Ecological Approach to Agricultural Development*, Santiago.

Raychaudhuri, S.P. *et al.*, 1963: *Soils of India,* New Delhi.

Singh, D.R. *et al.*, 2010: *Agro-climatic Atlas of Haryana,* HAU, Hisar.

Singh, J. 1976: *An Agricultural Geography of Haryana*, Vishal Publishing House, Kurukshetra.

Singh, J. and Dhillon, S.S., 1994: *Agricultural Geography,* Tata McGraw Hill, New Delhi.

Tewari, R.C., 2006: *Geography of India*, Praying Pustak Bhawan, Allahabad.

Thakur, B., 2011: *Depletion and Degradation of Land Resource of India*, R.S. Dube Foundation, Allahabad.

Trewatha, G.T., 1962: *The Earth's Problem Climate*, London.

Vaidyanathan, A., 2010: *Agricultural Growth in India: Role of Technology Incentives and Institutions,* Oxford University Press, New Delhi.

Wadia, D.N., 1960: *Geology of India*, London.

Wadia, D.N., and Auden, J.B. 1939: Geology and Structure of Northern India, *Memories of the Geological Survey of India*, Vol. LXXIII, Delhi.

5
Sources and Pattern of Irrigation

5.1 Introduction

Water is prime input to agriculture whether it is avilable through precipitation or irrigation because even the metabolic process will not begin without it what to say of vigorous vegetative growth and higher yield of crops. The rainfall the prime source of water, is periodic, seasonal and highly variable in its time of incidence, retreat, amount and its intensity. All these uncertainties in the occurrence of precipitation necessitate the use of irrigation water. Scientific and proper utilization of water resources for irrigation will enhance intensity of cropping and agricultural productivity (Narayanmoorthy, 2003). Therefore, for efficient utilization of the existing land capabilities and increasing the food resources, the existing rainfall situation has created acute need for irrigation.

Water is one of the primary bases and foundation of farming. Irrigation facilities are decisive factors in crop regionalization than any other determinant specially in the inherently fertile dry lands. Irrigation, being an artery and pulsating heart, is an absolute constraint as well as a sufficient volitional command over the location of commercial crops, improvement in the yield per hectare, swing in cropping pattern, change in the mechanism of landuse and dynamism in the livestock structure (Singh, 1976). Hence, the availability of adequate, timely and assured supply of water is an important determinant of agricultural productivity. Therefore, the development of high investment major and minor irrigation have to a quite an extent, freed agriculture from the dependence of monsoon rains.

Although India is blessed with abundant water resources, but the available water for irrigation purpose has been continuously diminishing. Agriculture alone consumes about 84 per cent of available water. About 69 million hectare metre of surface and 43.2 million hectare metre of groundwater is available for various purposes in India. The use of water for irrigation constitutes about 84 per cent of total utilization potential of water, i.e. 75 billion cubic metre. With the increase in demand of water for other uses, the share of water use for irrigation is likely to go down to nearly 73 per cent by 2025 A.D.

In Haryana—a water thirsty state, by and large, irrigation and power have been given top priority in the development plan, since, its inception so as to develop the agriculture at faster rate. Haryana has made commendable progress through irrigation facilities in laying the foundation of a progressive and forward looking agricultural activity. But Haryana is deficient in water potential. It has been estimated that if the entire cultivable area is provided irrigation facilities, sufficient for raring at least one crop, the total requirement of water for this purpose would be 20 million hectare metres, although the state is using only 11 million hectare metre water. With the adoption of Green Revolution which requires more irrigation water due to use of HYV of seeds and chemical fertilizers, the State Government has given high priority to utilize all surface and groundwater resources. New major and minor irrigation schemes were started and modernisation of old irrigation system was expedited. A scheme of the augmentation of deep tube-wells supplying groundwater to canals as well as tube-wells was also launched for the success of Green Revolution in the state. Therefore, it is imperative for the state to utilize and tap-up all possible water resources judiciously and to save as much water as possible by avoiding seepage, excessive and non-judicious irrigation. As a result, there has been rapid growth in the irrigated area in the state. The net irrigated area has increased from 20.85 million hectares in 1950-51 to 63-74 million hactares in 2008-09, while the gross irrigated area has increased from 22.6 million hactares to 88.87 million hactares during the same period.

However, there are considerable regional imbalances in the irrigation facilities and spatial differences in the trend of irrigation development that have swerved from the principle of equality. With areal differences in topography, water table, quality of groundwater etc., the irrigation facilities, the sources of irrigation and intensity of irrigation would naturally not be uniform both in space and time all over the state. Therefore, the distributional pattern is never exactly 'as two peas in a pod', The dry zone bordering the Thar desert covers about 13 per cent of the state and is prone to famine and droughts. Unfortunately, this region could not be covered by flow irrigation from any of the canal systems because of sandy undulations and dune concentration. The soil of this region is sandy and undulating, but the land is productive. Moreover, the groundwater is brackish and unfit for irrigation in most parts of the region. Therefore, there is a crying need for enhancing the irrigation facilities with appropriate water conservation and management practices in South-western Haryana.

In the recent past with the induction of minor irrigation development schemes such as sprinkler and drip irrigation, the irrigation facilities have considerably enhanced. In this region 2,87,900 hactares area amounting to nearly 68 per cent of the net sown area has been irrigated through various sources. Out of this, 18,200 hactares contributing nearly 65 per cent of the total irrigated area is irrigated through canals and the remaining 99,700 hactares amounting to nearly 35 per cent is irrigated with tube-well (Table 5.1). But there are considerable variations in the irrigated area ranging from as low as 19.5 per cent of net sown area in Siwani tahsil to as high and 90 per cent in Hisar tahsil. In Siwani, low percentage of irrigated area may be accounted to very low water table and lesser availability of canal irrigation water, while in Hisar availability of canal irrigation coupled with tube-well irrigation are the main reasons. Similarly, in Loharu (86.6%), Bhiwani (87.2%) and Dadri, (89.9%), the percentage of irrigated area is high due to above mentioned factors. In remaining tahsils of Tosham (42.5%), Narnaul (47.7%) and Mahendragarh (58.3), the percentage of irrigated area is low due to non-availability of both canal and tube-well irrigation.

Table 5.1: Spatial Pattern of Irrigation (2007-08)

Tehsil	*Area in ha*	*%age to*
Hisar	1,11,700	90.0
Bhiwani	35,000	87.2
Tosham	25,800	42.5
Loharu	38,200	86.6
Siwani	11,500	19.5
Dadri	53,100	89.9
Mahendragarh	7,400	58.3
Narnaul	5,200	47.4
Total	**2,87,900**	**67.8**

Note: The areal strength has been collected, adjusted and processed by the author based on the village records.

5.2 Sources of Irrigation

The above discussion represents aggregate picture of irrigated area in different tahsils, but there are many sources of irrigation in the region such as canals, tube-wells, wells, *rahat* etc. However, there are two major sources of irrigation in South-western Haryana, viz. canals and tube-wells. Very little area is irrigated with traditional method of irrigation such as *rahat*, well etc.

5.2.1 Canal Irrigation

In view of highly unfavourable climatic conditions, particularly low amount of rainfall and inadequate groundwater resources specially in south and south-west Haryana, the flow irrigation has a particular significance. As a result, Bhakra canal system, augmentation of canals, lift irrigation scheme were instituted which helped in increasing the irrigated area in the State, but the greatest beneficiaries were Hisar and Bhiwani districts. However, the importance of canal irrigation is relatively high in the central and south-western part of the State where it underlines the other sources of irrigation. However, canal is the major source of irrigation in the region irrigating nearly 64 per cent of the irrigated area (Fig. 5.1).

Table 5.2: Sources and Pattern of Irrigation (2005-06)

Tehsil	*Canals*		*Tube-wells*	
	Area in ha	*Per cent*	*Area in ha*	*Per cent*
Hisar	1,08,000	96-69	3,700	3.31
Bhiwani	29,500	84.28	5,500	15.72
Tosham	17,400	67.44	8,400	33.56
Loharu	11,600	30.37	26,600	69.63
Siwani	8,100	70.43	3,400	29.57
Dadri	13,400	25.24	39,700	74.76
Mahendragarh	100	1.35	7,300	98.65
Narnaul	100	1.92	5,100	98.08

Note: The area has been computed, adjusted and processed by the author based on the village records.

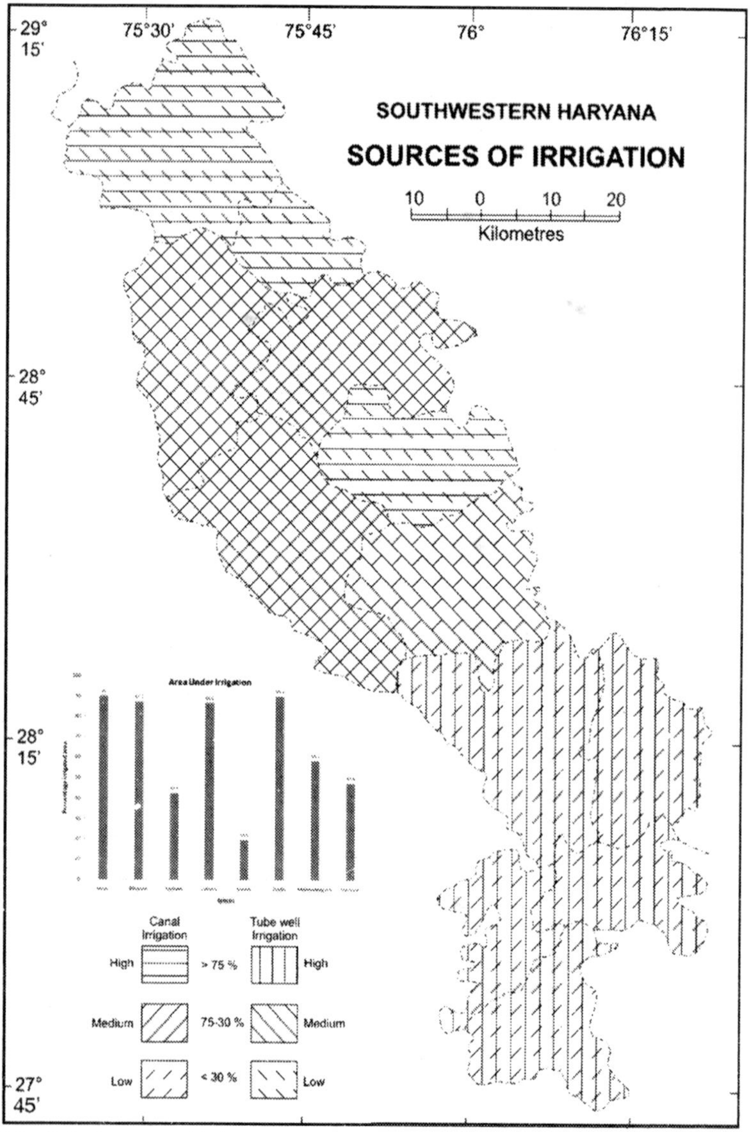

Fig. 5.1: Sources of Irrigation

Table 5.2 reveals very interesting pattern of irrigation. The northern part consisting of Hisar, Bhiwani, Tosham and Siwani have predominantly canal irrigation where nearly 97, 84, 67, and 70 per cent irrigation is through canals respectively. All these areas receive irrigation water from Western Yamuna and Bhakra canal systems. Besides these, the old canal systems also provide irrigating water. The intensity of canal irrigation decreases towards southern part of the region. Loharu and Dadri have little over one-fourth of their agricultural land irrigated by canals, while the extreme southern tahsils of Mahendragarh and Narnaul have less than 2 per cent of agricultural land irrigated through canals. The fundamental reason of this low percentage under canal irrigation is the undulating relief and larger distance of perennial source of water. In the northern part of the region, comparatively smooth relief, favourable slope and presence of perennial source of water are the fundamental causes of high percentage under canal irrigation which has brought comparatively more economic prosperity in this part of the region. The southern portion has the potentiality for extension of canal irrigation with the provision of lifting water from lower to higher areas.

5.2.2 Tube-well Irrigation

Tube-well irrigation which accounts nearly 35 per cent of the irrigated area in the region is second major source of irrigation. It is more popular in the southern part of the region than the northern. It accounts more than 98 per cent of the irrigated area in southern most two tahsils, viz. Mahendragarh and Narnaul. The basic reasons for such a high percentage under tube-well irrigation are undulating topography unfit for surface irrigation and the availability of good quality underground water. The areal coverage under tube-well irrigation decreases towards north which is just opposite to canal irrigation. Hisar has the lowest area under tube-well irrigation, while the tube-well irrigated area accounts nearly 30 per cent in Siwani, 33 per cent in Tosham, and 16 per cent in Bhiwani tahsils. Cheap and regular availability of canal water and deep underground water are the basic causes for lower areal coverage under tube-well irrigation.

The above analysis shows a lot of variations in the sources of irrigation. There are still much potentials for development of irrigation system in the region.

5.3 Sprinkler Irrigation

In spite of huge water potential in India, the demand for irrigation water will outstrip the supply potential to match the over quest of increasing food production to feed the teaming millions. World Trade Organisation (WTO) has given further incentive to the cultivators to incorporate water intensive crops in existing cropping pattern to enhance income and to meet-export demand. Moreover, the creation of intensive irrigation facilities and excessive use of canal water have aggravated the problem of water-logging, soil salinity, soil sodity etc., while groundwater is declining very fast due to over-exploitation of water resources through increased bore-wells and inefficient use of water due to adoption of age old practices in irrigation system. Therefore, selection of appropriate technology is of utmost importance for efficient utilization of scarce natural water.

Keeping above scenario in view, a number of demand managed marginal schemes, strategies and programmes have been introduced since late seventies in India to increase the water use efficiency. One of the demand managed strategies introduced relatively recently is micro-irrigation in which water is supplied at a required interval and quantity using pipe network, emitters and nozzles.

The conveyance and distribution losses are, therefore, reduced almost completely resulting into higher water use efficiency under minor irrigation, viz. Drip irrigation and sprinkler irrigation. There are distinct differences between the two methods in terms of flow rate, pressure requirement, wetted area and mobility (Kulkarni, 2005). The drip irrigation method supplies water directly to the root zone of the crops instead of land through a network of pipes with the help of emitters. As such the water losses occurring through distribution and evaporation are completely absent. (Narayanmoorthy, 1996 and Dhawan, 2003). In sprinkler irrigation method water sprinkles similar to rainfall into the air through nozzles which subsequently breaks into small water drops and fall on the field surface. Drip irrigation is mainly suitable and practiced for plantation crops, while the sprinkler irrigation is mainly for field crops. Hence, the water efficiency in drip irrigation is nearly 90 per cent, while it is only 30-40 per cent in sprinkler irrigation as it supplies water over the entire field of the crop (Kulkarmi, 2005).

Besides water use efficiency, micro-irrigation also delivers many other economic and social benefits to the society. It increases the productivity of different crops in the range of 20-90 per cent. It also reduces weed problems, soil erosion and cost of cultivation substantially specially in labour intensive operations. It also saves energy required for lifting water from wells.

5.3.1 Favourable Conditions for Sprinkler Irrigation

However, for the adoption and success of micro-irrigation method specially sprinkler irrigation, the most favourable areas are where:

- There is deficiency of both surface and groundwater resources.

- The surface method is not successful or possible due to undulating and rugged surface.
- There is high temperature bruising the crop due to lack of moisture.
- The soil has a complex structure ranging from sandy to stony.
- More area is to be irrigated with limited amount of irrigation water.
- There is lack of agricultural labourers.
- People have reasonable technical knowledge

From the above point of view South-western Haryana consisting of wholly or partially the community development blocks of Hisar-I, Hisar-II, Siwani, Tosham, Kairu, Badhra, Loharu, Kanina, Mahendragarh, Atehi Nangal, Narnaul and Nangal Chaudhry, has more favourable conditions for introducing sprinkler irrigation because this region has arid and semi-arid conditions which are characterized by lack of precipitation, high temperature, low humidly, poor availability of surface and ground-water, poor sandy, stony soil, and lack of labourer. It had been in the grip of chain of droughts due to scarce and irregular rainfall. The sandy and stony soils of this region make it unproductive. The ground-water being limited and brackish is unfit for irrigation in most part of the region. Therefore, the method of irrigation is an important aspect of better management of scantly water resources. And, sprinkler irrigation is one of such important methods.

5.3.2 Advantages of Sprinkler Irrigation

The sprinkler irrigation has many advantages over the other methods of irrigation including drip irrigation which is mostly suitable for plantation crops. However, it may be summarized as under:

- It helps in water conservation. In surface method 10-20 per cent water is soacked by the embankments. Moreover, more water is to be released in the field of uneven surface to reach its each corner, while in sprinkler irrigation water is uniformly sprinkled all over the field.
- The amount of water may be reduced or increased according to the requirement of the crop in its different stages of growth. As such in this method about 30-40 per cent water is saved and the irrigated area also increases by 50 per cent.
- It conserves soil as compared to surface irrigation in which the flow of water in undulating and rugged relief, erodes the land. In sprinkler irrigation, the erosion of soil does not take place and more areas are also brought under cultivation which is not possible in surface irrigation.
- Sprinkler irrigation keeps on washing the leaves of plants from dust which helps in absorption of sunlight and protects the crops from frost. It also washes away insects and pests from the leaves.
- It also saves labour cost which is spent in surface irrigation in making guls, embankments because there is no need of it in sprinkler irrigation.
- It is also helpful in spraying uniformly the chemical fertilizers and insecticides and pesticides as it is dissolved in water container and sprayed over the field.

5.3.3 Structure and Function of Sprinkler Irrigation Set

The sprinkler set mainly consists of a pumping set operated by

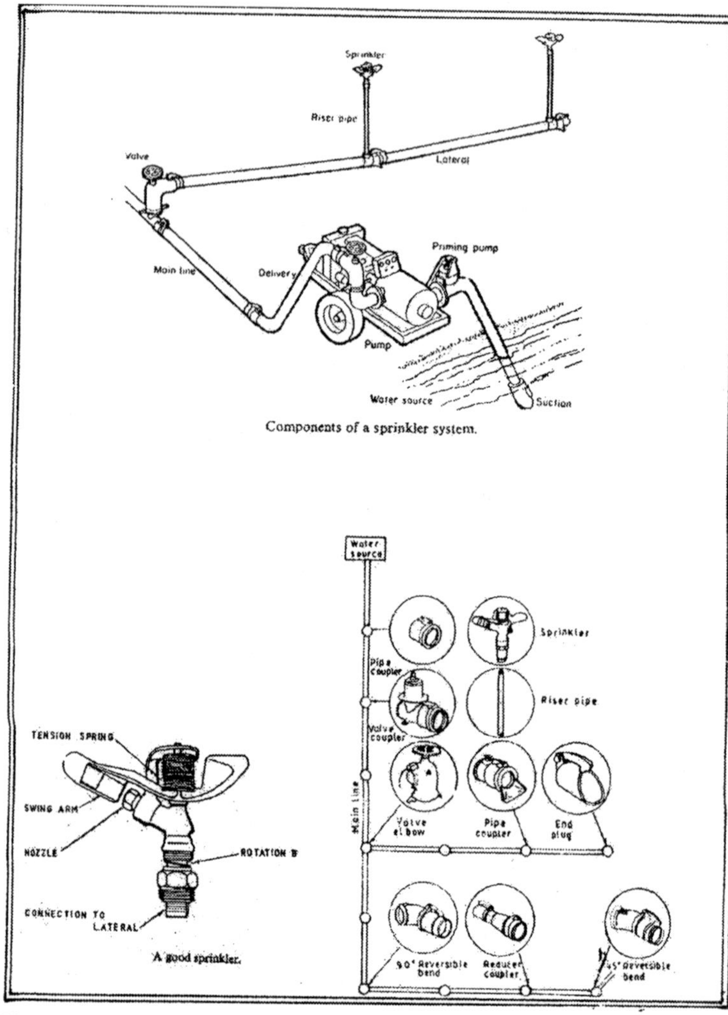

Components of a sprinkler system.

A good sprinkler.

Fig. 5.2: Model of Sprinkler Irrigation

electric motor or diesel engine, main pipe, lateral pipe, riser pipes, nozzle, coupler, band, reducer, foot button and end plug (Fig 5.2).

- Pumping set is used for lifting water either from tube-well or canal. It pumps water in the main pipe line.

- Pipes—main, lateral and riser are HDPI or aluminium pipes which are meant for distribution of water from the pumping station to the field.
- Nozzle is made of brass and is used for sprinkling the water coming from the riser pipe which stands about a foot above the surface level with the help of an adopter.
- 'T' and band are used for joining the additional lines or turning the direction of pipe.
- Pump connection nipple for connecting delivery line to sprinkler line.
- End plug is used for capping the last point of the pipe.

5.3.4 Fitting of Sprinkler Set in the Field

Sprinklers are fitted in the field in three ways, i.e. permanent, semi-permanent, and temporary. In permanent setting main pipe and lateral pipes are permanently laid down underground at a depth of nearly 1½ feet. In this method not only the pipes are saved and secure, but it saves labour for shifting the pipes time and again. In semi-permanent setting main pipeline is permanently laid down underneath but the lateral/branch lines are kept on the surface so that it can be removed from one site of the field to other site for irrigating the field. It is a more popular method of sprinkler irrigation, but in this method there is more labour cost and loss of irrigation water because the sprinkler sets are to be shifted from one site to the other site which takes about 20 minutes. In this duration either the motor is to be stopped or water is allowed to flow which goes waste. In temporary setting of sprinkler set, the entire setting is temporary. The set is shifted from one site to the other site as per requirement. It requires more labour cost in shifting the set and in the process the set is more prone to damage.

5.3.5 Sprinkler Irrigation in South-western Haryana

Sprinkler irrigation is quite popular method of irrigation in South-western Haryana. Government is giving huge amount of subsidies for the establishment of sprinkler irrigation set. A statistical account of number of sprinkler sets is given in Table 5.3, which indicates that there were 63,08,981 sprinkler sets in 2005-06, but they are not uniformly distributed across the region.

Since the sprinkler sets, by and large, are attached with tube-well, their distributional pattern is also in accordance with tube-well and well irrigation areas. Some sprinkler sets are also established in canals but their number is not very large.

Table 5.3 further shows that Mahendragarh alone has 18,557 sprinkler sets which accounts early 30 per cent of the total sets, while Narnaul has 10,456 sprinkler sets, which accounts nearly 17 per cent of the total number.

Table 5.3: Tahsil-wise No. of Sprinkler Sets (2005-06)

Tahsils	*No. of S.S.*	*%age to total*
Hisar	7,210	11.43
Bhiwani	4,145	6.57
Tosham	5,917	9.38
Loharu	8,134	12.89
Siwani	3,990	6.33
Dadri	4,677	7.41
Mahendragarh	18,557	29.41
Narnaul	10,456	16.58
Total	**63,089**	**100.00**

The rugged undulating topography, sandy soil and lack of both surface and groundwater are the main reasons for higher number of sprinkler sets in these tahsils. The number of sprinkler sets decreases from south towards north and the lowest number is in Bhiwani tahsil followed by the tahsil of Siwani. There are 4,145 sprinkler sets in Bhiwani and 3990 sprinkler sets in Siwani amounting to 6.6 per cent and 6.33 per cent respectively. In the remaining tahsils, the percentage contribution is more than one-tenth of the total in each tahsil. In the northern region the dominance of canal irrigation is the main cause for lesser number of sprinkler sets.

5.3.6 Intensity of Sprinkler Sets

The absolute number of sprinkler sets does not give the clear and precise picture of its status due to varying sizes of the tahsils and differential percentage of the cultivated land. Therefore, for the sake of comparison of the number of sprinkler irrigation sets per 100 hectares of net sown area has been worked out and is given in Table 5.4.

Table 5.4: No. of Sprinkler Sets per 100 ha of Net Sown Area (2005-06)

Tahsils	*No. of sets*	*N.S.A.*	*No. S.S. 100 Ha of NSA*
Hisar	7,213	1,23,933	5.82
Bhiwani	4,145	20,100	20.62
Tosham	5,917	60,700	9.75
Loharu	8,134	58,900	13.81
Siwani	3,990	44,100	9.05
Dadri	3,990	44,100	9.05
Mahendragarh	18,557	78,000	23.79
Narnaul	10,456	71,300	14.66

Note: The data have been collected, streamlined and processed on the basis of the village records by the author.

Table 5.4 shows a lot of variations in the availability of number of sprinkler sets per hundred hectares of net sown area. It varies from as low as 5.82 in Hisar to as high as 23.79 sprinkler sets per 100 ha of net sown area in Mahendragarh tahsil. The reasons in both the cases are the same as given in case of number of sprinkler sets. Its distributional pattern is also in accordance with tube-well irrigation system. There are two tahsils of Mahendragarh and Bhiwani where number of sprinkler sets per 100 hectares of net sown area is more than 20, while in Loharu (13.81) and Narnaul (14.66), it is between 10-20 sprinkler sets per 100 ha of net sown area. In the remaining 4 tahsils viz. Tosham (9.75), Dadri (7.91), Siwani (9.05) and Hisar (5.82) number of sprinkler set is less than 10 per 100 ha of net sown area. The irrigational pattern with sprinkler method is almost the same as in case of tube-well irrigation. And, therefore, there is no need to repeat it again.

References

Aggarwal, M.C. *et al.*, 1982: Yield of *Bajra* and Wheat with Saline Water Applied Through Sprinkler and Surface Irrigation Methods, *Annals of Arid Zones*, Vol. 21, No. 1.

Agricultural Statistics at a Glance, 2006-2012. Directorate of Economics and Statistics, Ministry of Agriculture, New Delhi.

Anonymous, 1998: Sprinkler Irrigation in India, *INCID*, New Delhi.

Clark, *et al.*, 2000: Micro-Irrigation Design Consideration for Sand Soil Vegetation Production System, *International Water and Irrigation Review*, Vol. 19, No. 4.

Dash, N.K., 1998: *Drip Irrigation and Micro-Irrigation Potential and Prosperity for Sugarcane Cultivation with Reference to Orissa*, VSI.

Desai, N.D. *et al.*, 1993: Feasibility of Mini sprinkler in Onion, in *EBIP*.

Dhawan, B.D., 1988: Role of Irrigation in Raising Intensity of Cropping, *Journal of Indian School of Political Economics*, Vol. 3, No. 4.

Dhawan, B.D., 2002: *Technological Change in Indian Irrigated Agriculture: A Study of Water Saving Method*, Commonwealth Publishers, New Delhi.

Government of India, 1999: *Report of Working Group on Water Availability for Use,* National Commission for Integrated Water Resource Development Plan, New Delhi.

Hepase *et al.*, 1993: Micro-Irrigation in Sugarcane Cultivation., *CBIP*.

Hooda, R.S. 1994: Role of Capital on Adaptation of Sprinkler Irrigation at Different Size of Holdings in Bhiwani District, *Crop Research* Vol. VIII, No. 2.

Hooda, R.S. 1994: Impact of Sprinkler Irrigation on Crop-combination. A Case Study of Bhiwani District, Haryana, *Crop Research,* Vol. 7, No. 1.

Hooda, R.S. 1996: Diffusion of Sprinkler Irrigation in Arid India, *Crop Research,* Vol. 11, No. 2.

IMCID, 1998: *Sprinkler Irrigation in India,* New Delhi.

Indian Agricultural Statistics, Vols. I & II, Directorate of Economics and Statistics, Government of India, New Delhi.

Jain, S.K., Pushpendra, A.K. and Vijay, P.S., 2007: *Hydrology and Water Resources of India.*

Menon, R.A., 1995: Micro-Irrigation: Profits in Pipeline, *India Today*, May.

Mohammad, N., and S.C. Rai, 2014: *Agricultural Diversification and Food Security in Uttarakhand*, Concept Publishing Co. (P) Ltd., New Delhi.

Mohanmad, N., 1996: *New Dimensions in Agricultural Geography*, Vols. I-VII, Concept Publishing Co. (P) Ltd., New Delhi.

Narayanmoorthy, A. and Deshpande, R.S., 2003: Irrigation Development and Agricultural Wages: An Analysis Across States, *EPW*, Vol. 38, No. 35, Delhi.

Narayanmoorthy, A., 2004: Drip Irrigation in India: Can it Solve Water Scarcity, *Water Policy*, Vol. VI, No. 2.

Patra, P., 2006: *Land Degradation in Tropical Highlands*, Rajesh Pub., New Delhi.

Rao, C.H., 2002: "Sustainable use of Water for Irrigation in Indian Agriculture" in Raj Kapila and Unna Kapila (eds.) *Economic Development in India*, Vol. 59, New Delhi.

Sanklayan, P.L. and Singh, I.P., 1985: Impact of Surface and Lift Irrigation System on Cropping Pattern, *Economic Affairs,* India.

Seckler, D., Banker, R. and Amar Singh, U.A., 1999: *International Journal of Water Resources Development,* Vol. 15, No. 2.

Shafi, M., 2006: *Agricultural Geography*, Pearson Pub., New Delhi.

Singh, H.P., *et al.*, 2001: *Micro-Irrigation, CBIP*, Delhi.

Singh, J. and Dillon, S.S., 1984: *Agricultural Geography*, Tata McGraw Hill Pub., New Delhi.

Singh, J., 1976: *An Agricultural Geography of Haryana.* Vishal Pub. Kurukshetra.

Sivanappan, R.K., 1994: Prospects of Micro Irrigation in India, *Irrigation and Drainage System*, Vol. VIII, No. 1.

Srivastava, G.S., Singh, I.B. and Kulshreshtha, 2006: Late Quarternary Geomorphic Evaluation of Yamuna Sutlej Interfluvens: Significance and Terminal Fan, *Journal of Indian Society of Remote Sensing*.

Verma, C.J.V. and Rao, A.R.G. (eds.), 1998: Micro-Irrigation and Sprinkler Irrigation System, *CBIP*, New Delhi.

6
Spatial Pattern of Landuse/ Land Cover

6.1 Introduction

The patterns of landuse/land cover is the function of physio-socio-economic conditions of an area. It is more so in a semi-arid desert ecosystem like South-western Haryana where all these conditions are highly varied, vulnerable, fragile and complicated. There are relict mountains and its out-crops, vast expanse of sand covered with sands, sand dunes, inter-dunal plains, old alluvial plains, closed inland drainage system, highly varied and unpredictable climate, infertile sandy soil, low level of socio-economic conditions and lower level of technology. All these varied geographical conditions have resulted into complex pattern of landuse/land cover which needs to be examined for the formulation of a rational landuse plan and policy for agricultural development to meet the growing demand of the region (Singh and Dhillon, 1984 and Shafi, 2006). It should be properly utilized, managed and conserved using modern tools and techniques (Mohammad, Singh and Datta, 2007).

6.2 Spatial Pattern of Landuse/Land Cover

Since the landuse/land cover model developed by Remote Sensing

Agency, Hyderabad has not been fully adopted across the country, the model developed by Technical Committee on Co-ordination of Agricultural Statistics has been used here to describe and explain the landuse/land cover in this study. It is very much affected by the surface terrain and climatic conditions. A statistical account of landuse/land cover for the year 2005-06 is represented in Table 6.1.

Table 6.1 shows that more than 82 per cent of the total area is under cultivation and less than 18 per cent is under rest of the categories of landuse. It clearly indicates the predominance of agriculture and heavy dependence on it. A brief account of each category is presented in Fig. 6.1.

Table 6.1: Spatial Pattern of Landuse/Land Cover (Area in Ha, 2005-06)

Tahsil	*Forest*	*Land not available*	*Other Uncultivated land*	*Fallow land*	*Net Sown Area*
Hisar	67 (0.05)	16,800 (11.36)	—	7,200 (4.84)	1,23,933 (83.75)
Bhiwani	75 (0.30)	2,650 (10.62)	—	2,125 (8.51)	20,100 (8057)
Tosham	700 (0.93)	6,500 (8.68)	—	7,000 (9.35)	60,700 (81.04)
Loharu	600 (0.88)	7,200 (10.54)	—	1,600 (2.34)	57,800 (86.24)
Siwani	700 (1.43)	3,300 (6.76)	—	700 (1.43)	44,100 (90.38)
Dadri	200 (0.29)	6,850 (9.99)	—	2,400 (3.50)	59,100 (86.21)
Mahendragarh	1,800 (1.82)	16,500 (10.72)	—	2,400 (2.43)	7,800 (79.03)
Narnaul	300 (0.31)	19,600 (19.50)	1,400 (0.22)	3,800 (3.95)	71,300 (74.74)
Total	**4,442** **(0.71)**	**78,400** **(12.49)**	**1,400** **(0.22)**	**27,225** **(4.34)**	**5,16,133** **(82.24)**

Notes: 1. The data have been collected, streamlined and processed on the basis of village records by the author.

2. Figures in parentheses indicate percentage to total area.

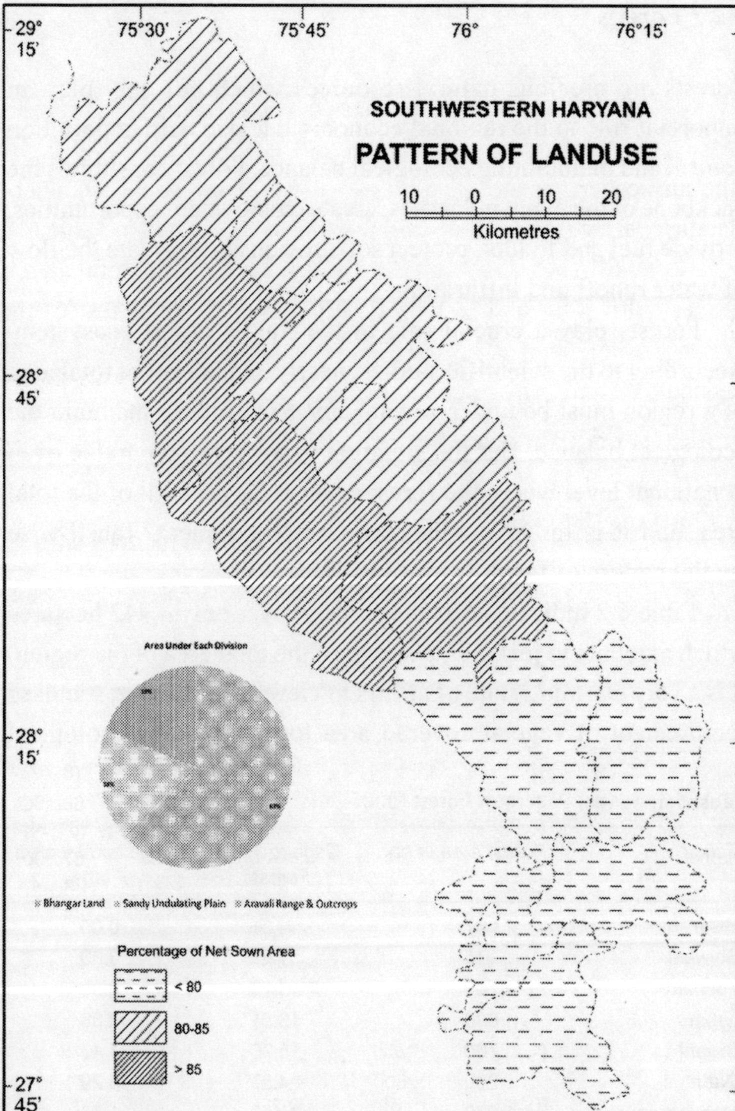

Fig. 6.1: Patterns of Landuse 2005-06

6.2.1 Forests

Forests are precious natural resources which not only play an important role in the regional economy but also help in pollution control and maintaining ecological balance. Forest constitutes the backbone of the many industries, create employment opportunities, provide fuel and fodder, protect soil erosion and regulate the flow of water runoff and infiltration.

Forests play a crucial role in the fragile desert ecosystem. According to the scientific studies, nearly 33 per cent of total area of a region must be under vegetal cover in order to maintain the ecological balance. But the situation is not very favourable even at national level where the forest cover is 22 per cent of the total area, and it is far below the standard requirement. Tahsil-wise spatial pattern of forest is given in Table 6.2.

Table 6.2 indicates that the forests cover only 4,442 hectares which amounts to just 0.71 per cent of the total area of the region. It is a very alarming state of affairs in view of 33 per cent standard requirement of vegetal covered area for maintaining ecological

Table 6.2: Spatial Pattern of Forest (2005-06)

Tahsil	*Forest area in ha*	*%age to Total Forest*	*%age to total area of tahsil*
Hisar	67	1.51	0.05
Bhiwani	75	1.69	0.30
Tosham	700	15.76	0.93
Loharu	600	13.51	0.88
Siwani	700	15.76	1.43
Dadri	200	4.50	0.29
Mahendragarh	1,800	40.52	1.82
Narnaul	300	6.75	0.31
Total	**4,442**	**100.00**	—

Note: The data have been collected, streamlined and processed on the basis of village records and information by the author.

balance. Tahsil-wise spatial pattern is quite uneven. The forest coverage in the tahsils ranges from as low as 0.05 per cent in Hisar tahsil to as high as 1.82 per cent in Mahendragarh.

Leaving the two tahsils, viz. Mahendragarh (1.82%) and Siwani (1.43%), the forest cover in all the tahsils is not even two per cent of the total area of the tahsil. Therefore, there is an urgent need to devise means and strategies to increase the forest cover urgently to save the fragile desert ecosystem. Likewise, the percentage of forest cover in different tahsils to the total forest cover in the region is quite uneven ranging from 1.51 per cent (lowest) in Hisar to 40.52 per cent (highest) in Mahendragarh. The highest percentage in Mahendragarh could be accounted to the rocky hiliy areas and comparatively higher amount of rainfall. Siwani (15.76%), Tosham (15.76%) and Loharu (13.51%) account nearly 15 per cent or more to the total forest cover in the region.

6.2.2 Land Not Available for Cultivation

This category includes two types of landuse: (1) area under non-agricultural uses such as settlements, roads, canals, ponds etc., and (ii) rocky, barren uncultivated land including hills, plateau and mountainous areas. The spatial pattern of this category is also highly uneven ranging from nearly 3.3 per cent (the lowest) in Bhiwani to 23.7 per cent (the highest) in Narnaul tahsil, where ranges of Aravali with barren hills and exposed rocks are found. The spatial pattern of this category is represented in Table 6.3 which shows that in three tahsils, i.e. Hisar, Mahendragarh and Narnaul, it accounts more than one-fifth of the total area under this category in the region. The other three tahsils—Tosham, Loharu and Siwani—account nearly 10 per cent of the total area. In the remaining two tahsils, i.e. Bhiwani and Dadri, it covers nearly 4 and 3 per cent respectively. This percentage variation range may be explained in terms of desert ecosystem, Aravali ranges, and uneven spatial pattern of population.

Table 6.3: Spatial Pattern of Land not available for Cultivation (2005-06)

Tahsil	*Area in ha*	*%age to total under this category*	*%age to total area of Tahsil*
Hisar	16,800	21.43	11.36
Bhiwani	2,650	3.38	10.62
Tosham	6,500	8.29	8.68
Loharu	6,850	8.75	9.99
Siwani	7,200	9.18	10.54
Dadri	3,300	4.21	6.76
Mahendragarh	16,500	21.05	16.72
Narnaul	18,600	23.71	19.50
Total	**78,400**	**100.00**	—

Note: The data have been collected, stream lined and processed on the basis of village records by the author.

The percentage contribution of this category to the total area of the respective tahsils is not very uneven. It accounts almost 10 or more than 10 per cent of the total area of the tahsil. The highest coverage is in Narnaul tahsil where it accounts nearly one-fifth of the total area, while the lowest (6.76%) is recorded in Siwani tahsil.

6.2.3 Other Uncultivated Land Excluding Fallow Land

This category includes three sub-categorics, viz. (i) permanent pasture and grazing land, (ii) land under miscellaneous three crops and groves, and (iii) culturable waste land. The areal coverage of this category is almost neglible/absent except in Narnaul tahsil where it covers 1,400 hectares or 1.47 per cent of the total area. The conspicuous absence of this category clearly indicates that all culturable land have been brought under cultivation due to expansion of irrigation facilities and other parameters of Green Revolution.

6.2.4 Fallow Land

This is a type of land which is left uncultivated for more than one year, so that, it could recoup its fertility lost due to excessive cultivation or is comparatively less fertile to produce crops continuously. There are three types of fallow land, viz., seasonal fallow which refers to a type of land left fallow in one or the other agricultural season in rotation system of crops and it is included in net sown area; current fallow which is left unutilized for one or two successive years; and old fallow which is uncultivated for 5 to 7 years to recoup its fertility. If the agricultural land is left fallow beyond this period, it is termed as culturable wasteland. The number of years for each category is not constant and varies from state to state.

The fallow land accounts about 4 per cent of total area of the region, but its distributional pattern is also quite uneven which is represented in Table 6.4.

The highest hectare of fallow land amounting to 7,200 hectare contributing more than 26 per cent of this category is found in

Table 6.4: Spatial Pattern of Fallow Land (2005-06)

Tahsil	*Area in ha*	*%age total under the category*	*%age to total area of tahsil*
Hisar	7,200	26.45	4.84
Bhiwani	2,125	7.80	8.51
Tosham	7,000	25.71	9.35
Loharu	2,400	8.82	3.50
Siwani	1,600	5.88	2.34
Dadri	700	2.57	1.43
Mahendragarh	2,400	8.82	2.43
Narnaul	3,800	13.95	3.98
Total	**27,225**	**100.00**	—

Note: The data have been collected, stream lined and processed on the basis of village records by the author.

Hisar followed by its southern adjacent tahsil Tosham, where this category covers more than one-fourth of fallow lands. Narnaul, the southern most tahsil, has 3,800 hectare fallow land which amounts to nearly 14 per cent of fallow land. In Dadri tahsil, where the soil is comparatively more fertile, the area under fallow land is lowest, i.e. 700 hectares which constitutes just 2.57 per cent of the fallow lands. In the remaining 3 tahsils, the percentage share of fallow land is around 8 per cent. This category has the maximum potentiality to be brought under cultivation by improving the agricultural infrastructure.

6.2.5 Net Sown Area

This is the real agricultural land on which the cultivators perform all their agricultural activities to get their livelihood. The cultivators always try their level best to bring more and more land under cultivation. In the study region they have been quite successful due to adoption of agricultural technology. More than 82 per cent of the total area is under cultivation, but its distributional pattern

Table 6.5: Spatial Pattern of Net Sown Area (2005-06)

Tahsil	*Area in ha*	*%age total area under this category*	*%age to total area of the Tahsil*
Hisar	1,23,933	24.01	83.75
Bhiwani	20,100	3.89	80.57
Tosham	60,700	11.77	81.04
Loharu	58,900	11.45	86.24
Siwani	41,100	11.41	90.38
Dadri	59,100	8.54	86.21
Mahendragarh	78,000	15.11	79.03
Narnaul	71,300	13.82	74.74
Total	**51,61,133**	**100.00**	—

Note: The data have been collected, streamlined and processed on the basis of village records and information by the author.

is also uneven ranging from nearly 75 per cent in Narnaul to about 91 per cent in Siwani tahsil. Its tahsil-wise spatial pattern is given in Table 6.5. and Fig. 6.1.

It is evident from Table 6.5 that except Narnaul tahsil, net sown area covers more than 80 per cent of the total area in all tahsils. In Narnaul tahsil, its percentage is low due to Aravali hills and its outcrops. In Dadri the percentage of net sown area is high due to comparatively more fertile soil and higher amount of rainfall. However, there is very little scope of further expansion of net sown area. Therefore, for increasing the cropped area, the intensity of cropping has to be increased.

6.3 Intensity of Cropping

Intensity of cropping is a measure to express the use of land during one agricultural year. It is mainly influenced by ecological, techno-economic and socio-cultural factors. By and large, the land is put to more intensive use in the areas where agricultural land is scarce and its demand is more. In such areas land is put to maximum use at the given level of technology. This system gradually deteriorates the soil and the yield becomes low. It is, by and large, practiced in the developing countries with high density of population. On the contrary, almost opposite is the case where land is in abundance and its demand is low due to low density of population, and the large population being engaged in secondary and tertiary activities. In this system the yield is high and the level of soil deterioration is low. This system in more prevalent in the developed countries with lower density of population.

Since, the crops are being produced on the net sown area, the intensity of cropping is measured with reference to it. There are many ways of measuring the intensity of cropping but the most common method of its measurement is given below:

$$I.C. = \frac{GCA}{NA} \times 100$$

where

$I.C.$ = Intensity of cropping;

GCA = Gross cropped area; and

NSA = Net sown area.

Following this method the intensity of cropping has been measured in South-western Haryana and the results are presented in Table 6.6.

Table 6.6 shows that the net sown area is very intensively utilized in spite of adverse physical environment. The average intensity of cropping of the whole region is 188 per cent, although there are tahsils like Bhiwani and Tosham where the intensity of cropping is more than 205 per cent. It means that some of the net sown area in these tahsils has been sown three times in one agricultural year. It is mainly due to timely availability of irrigation water in adequate quantity and use of modern agricultural infrastructure.

Table 6.6: Spatial Pattern of Intensity of Cropping: 2005-06

(Area in Hectares)

Tahsil	*Net sown area*	*Gross cropped area*	*Intensity of cropping*
Hisar	1,23,933	2,05,056	165.00
Bhiwani	20,100	4,11,314	205.50
Tosham	60,700	1,24,900	205.76
Loharu	58,900	1,17,613	198.37
Siwani	41,100	1,16,700	198.37
Dadri	59,100	86,600	197.48
Mahendragarh	78,000	1,50,308	192.70
Narnaul	71,300	1,30,009	182.34
Total	**51,61,133**	**9,72,500**	**188.00**

Note: The data have been collected, streamlined and processed on the basis of village records by the author.

In most of the remaining tahsils it is almost approaching to 200 per cent, e.g. it is 198, 197.48, 196.37, and 192.70 per cent in Loharu, Dadri, Siwani and Mahendragarh tahsils respectively. There are only two tahsils viz. Hisar and Narnaul where intensity of cropping is 165 and 182.34 per cent respectively.

In the light of the above discussion, it may be inferred that the horizontal and vertical expansion of agriculture in South-western Haryana has almost reached a saturation point. Therefore, the scope of increasing agricultural production lies in enhancing the yield per unit of agricultural land.

References

Brerno, Dorin and Frederi, L. (eds.), 2009: *Agriculture and Food in India*, Manohar Pub., New Delhi.

Department of Economics and Statistical Analysis, 2006:
(i) *Economical Survey of Haryana.*
(ii) *Statistical Abstract of Haryana, 2008 to 2006.*

Director of Economics & Statistics: *Indian Agricultural Statistics,* Vols. I & II, New Delhi.

Director of Economics and Statistics, *Agricultural Statistics at a Glance*, 2006 to 2012, New Delhi.

Duggal, S.L., 1975: *Soil Geography of Haryana.*

Gajbheye, K.S. and Mandal, C., 2000: Agro-ecological Zones: Their Soil Resources and Cropping System. *Status of Farm Management in India*, New Delhi.

Mohammad, N., 1992: *New Dimensions in Agricultural Geography*, Vols. I-VIII, Concept Publishing Co., New Delhi.

Mohammad, N. Singh, R.B., and Datta, A. (eds.) 2007: *Spatial Information Technology for Natural Resource Management*, Concept Publishing Co., New Delhi.

Mohammad, N., 1978: *Agricultural Landuse in India*, Inter-India Publication, New Delhi.

Shafi, M., 2006: *Agricultural Geography*, Pearson Pub., New Delhi.

Singh, D.R. *et al.*, 2010: *Climatic Atlas of Haryana,* Hisar.

Singh, J. and Dhillon, S.S., 1984: *Agricultural Geography,* Tata McGraw Hills, New Delhi.

Singh, R.P., 1992: 'Concept of Land' in Mohammad, N. (ed.) *New Dimensions in Agricultural Geography*, Concept Publishing Co., New Delhi.

Singh, J., 1974: *An Agricultural Atlas of India: A Geographical Analysis*, Vishal Pub., Kurukshetra.

7
Spatial Pattern of Agricultural Landuse and Level of its Development

7.1 Introduction

Since all geographical factors—physio-ecological, techno-economic and socio-cultural—are highly varied and complex at all levels from macro- to micro—both in space and time, on the one hand, and the requirements of these geographical factors for all species of agricultural plants right from the process of metabolism to the ripening of crops varies, on the other hand, there are a lot of variations in the cultivation of crops across the region (Mohammad, 1972, and Ahmad, 2004). Variety of crops are grown in this region ranging from cereal crops to commercial crops. However, there are two main agricultural seasons in the region, besides Zaid season—a season of additional crops in which a very small percentage of net sown area is devoted to cultivation of summer vegetables, watermelon, and some fruit crops. The principal crops of *kharif* season are *guar*, bajra, cotton and some pulses, while main crops of *rabi* season are wheat, barley, gram, oilseeds, and pulses (Table 7.1). Major emphasis is given on cultivation of foodgrains which cover 7,08,600 ha accounting nearly 72.86 per cent of gross cropped area. Oilseeds have covered

1,24,800 ha accounting nearly 13 per cent of the gross cropped area.

Table 7.1: Area Under Major Crops (2005-06)

Crops	*Area in ha*	*Percentage to*	
		Net sown area	*Gross cropped area*
Rice	8,300	1.61	0.85
Bajra	2,98,800	57.89	30.72
Wheat	1,50,100	29.08	15.43
Barley	70,000	1.36	0.72
Gram	90,200	17.48	9.28
Other Pulses	91,200	24.17	9.38
Oil seeds	1,24,800	13.87	12.84
Cotton	71,600	7.36	7.36
Others	1,30,500	25.28	13.42
Total	**10,35,500**	—	**100.00**

Note: Data have been collected, streamlined and processed on the basis of village records by the author.

Cotton—a commercial crop covers 71,600 ha or about 7 per cent of gross cropped area. There are several other crops like lentil, cowpeas, *guar*, etc. which cover a small portion of gross cropped area, but they put together cover 1,30,500 ha accounting nearly 13 per cent of the gross cropped area.

Above discussion gives an aggregate picture of the areal strength of different crops in the region during both the agricultural seasons, but the clear picture will emerge when we look into its spatial pattern which is given in Table 7.2 and Fig. 7.1.

7.2 Spatial Pattern of Agricultural Landuse

Table 7.2 shows the areal strength of the major crops in each of the 8 tahsils of the region. It further shows that *bajra* which covers 29,880 ha amounting to nearly 31 per cent of the gross cropped area is the major crop of the region. It is followed by wheat and

Table 7.2: Spatial Pattern of Agricultural Landuse (2005-06)

(Area is in 100 ha)

Tahsils	*Rice*	*Bajra*	*Wheat*	*Barley*	*Gram*	*Other Pulses*	*Oilseeds*	*Cotton*
Hisar	48	218	692	33	68	133	197	464
Bhiwani	17	80	77	3	17	17	99	45
Tosham	—	329	106	4	224	234	266	70
Loharu	—	276	197	7	33	38	346	70
Siwani	—	752	182	14	116	13	278	34
Dadri	18	219	36	1	334	27	62	29
Mahendragarh	—	594	276	6	34	NA	NA	3
Narnaul	—	520	184	2	76	NA	NA	1
Total	**83**	**2,988**	**1,750**	**70**	**902**	**912**	**1248**	**716**

Note: Data have been collected, streamlined and processed on the basis of village records by the author.

N.A. = Not available.

oilseeds which occupy 15,010 and 12,480 ha respectively. However, it would be better if a brief account of each major crops is given here.

7.2.1 Bajra (Pennisetum Typhoides)

Bajra is the most important millet crop of the region and it occupies the prime position. It is also known as pearl millet, bulrish millet and spike millet as at the head of the spike is a compact concentration of grey grain. It is a high protein heat giving grain suffering less from environmental hazards than other food grains. *Bajra* is successfully grown in warm areas having low rainfall because it is a quick growing crop. Intermittent light shower, bright sunshine, low humidity and considerable diurnal range of temperature during growth period, are ideal conditions for its cultivation. It is a heat and drought tolerant crop. Heavy rains are disastrous for this crop as the pollen is washed away and many of the female flowers are not fertilized which ultimately reduces the yield of crop. It grows best on the shallow well-drained light loamy

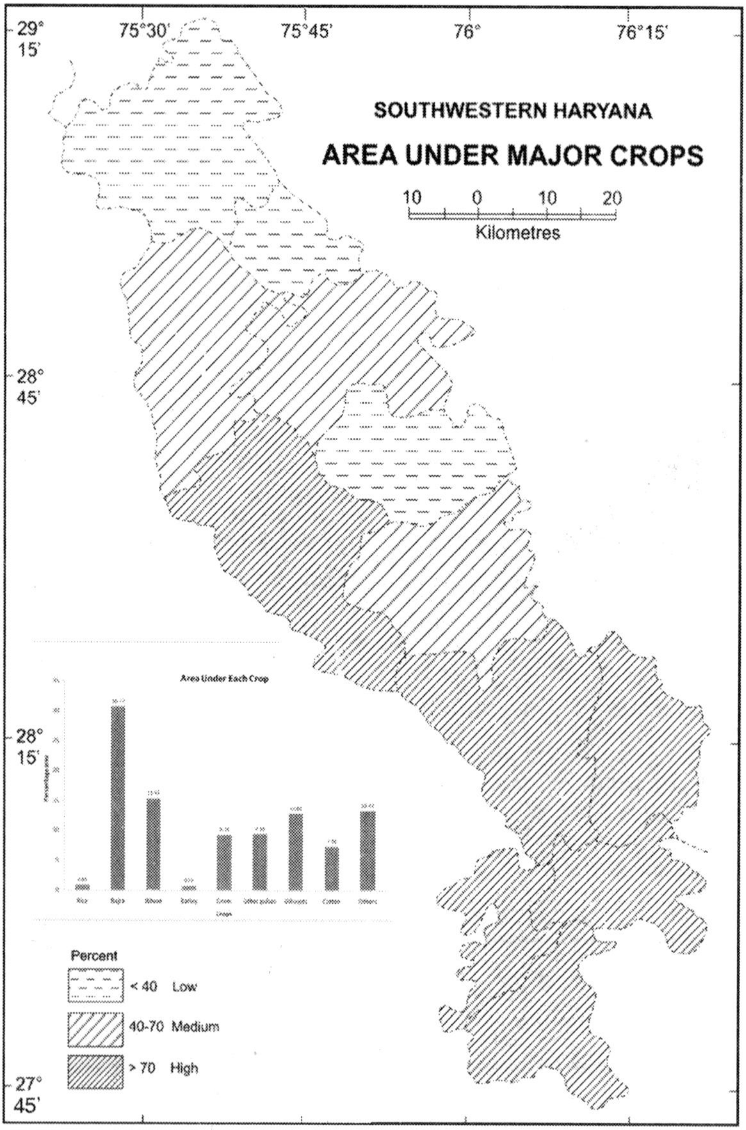

Fig. 7.1: Area under Major Crops 2005-06

or sandy soils which are poor soils lacking in humus. As such its concentration is the reflection of the distribution of rain, the length of rainy season and soil. *Bajra* is widely grown all over the region. It covers 2,98,800 ha accounting nearly 31 per cent of the gross cropped area or 57.89 per cent of net sown area. It is a *kharif* crop sown just after the onset of the monsoon season and is harvested in the month of September-October. A statistical account of the spatial pattern of *Bajra* is given in Table 7.3 and Fig. 7.1.

Table 7.3 shows that *bajra* is a prominent crop in Loharu tahsil where it occupies 75,200 ha accounting nearly 128 per cent of the net sown area which shows that more than one-fourth of net sown area produces two crops of *bajra* in one agricultural season. It also covers nearly half of the gross cropped area in the tahsil of Loharu. Mahendragarh and Narnaul are other two tahsils where it covers 59,400 and 52,000 ha covering nearly 76 and 73 per cent of the net sown area respectively. It also occupies nearly two-fifths of the gross cropped area in each of the tahsils.

Table 7.3: Spatial Pattern of Bajra (2005-06)

Tahsils	*Area in ha*	*Percentage to*	
		Net sown area	*Gross cropped area*
Hisar	21,800	17.59	10.63
Bhiwani	8,000	39.80	19.31
Tosham	32,900	54.20	26.34
Loharu	75,200	127.67	49.31
Siwani	21,900	49.66	25.29
Dadri	27,600	46.70	23.47
Mahendragarh	59,400	76.15	39.52
Narnaul	52,000	72.93	40.00
Total	**2,98,800**	**57.89**	**30.72**

Note: Data have been collected, streamlined and processed on the basis of village records by the author.

In fact, Loharu, Mahendragarh and Narnaul lying in south and south-western part of the region form a contiguous belt of high concentration of *bajra* cultivation covering more than 40 per cent of gross cropped area. Moderate to high temperature, coarse sandy soil, scanty rainfall but reasonable irrigation facilities are favourable conditions for its concentration. The second region with 20-40 per cent of gross cropped area constitutes Siwani, Tosham and Dadri tahsils. The third zone with less than 20 per cent coverage of gross cropped area constitutes Hisar and Bhiwani tahsils. It is clear from the above discussion that the concentration of *bajra* cultivation is in the south and south-west and it decreases towards north and north-east.

7.2.2 Wheat (Tritium Aestivum)

Wheat is one of the most important small grain cereal crops in the region. Its cultivation is largely determined by agronomic, social and economic conditions. It is grown in a wide range of climatic and adaphic conditions ranging from dry sub-humid to semi-arid climatic condition, and stiff clay to sandy loam adaphic status. It requires sufficient moisture for its unchecked growth. This crop is very sensitive to soil moisture deficit during winter which results into requirement of irrigation water. Wheat is a rabi crop sown in the month of November-December and is harvested in March-April. Wheat is next important crop in terms of its areal strength. Spatial pattern of wheat is given in Table 7.4.

It is obvious from Table 7.4 that wheat occupies about 29 per cent and 15 per cent of the net sown area and gross cropped area in the region respectively, but there are a lot of differences in percentage strength of wheat with reference to both net sown area and gross cropped area. In Hisar wheat occupies nearly 56 per cent of net sown area. However, with the exception of Tosham

Table 7.4: Spatial Pattern of Wheat (2005-06)

Tahsil	*Area in ha*	*Percentage to total*	
		Net sown area	*Gross cropped area*
Hisar	69,200	55.84	33.73
Bhiwani	7,700	38.31	18.64
Tosham	10,600	17.46	8.49
Loharu	18,200	30.89	11.93
Siwani	3,600	8.16	4.16
Dadri	19,700	33.33	16.75
Mahendragarh	27,600	38.71	18.36
Narnaul	18,400	25.81	14.15
Total	**17,500**	**29.08**	**15.43**

Note: Data have been collected, streamlined and processed on the basis of village records by the author.

which breaks the continuity, a compact zone with more than 30 per cent of net sown area under wheat, is formed by Hisar, Bhiwani, Dadri and Mahendragarh tahsils. The lowest percentage of wheat is recorded in Siwani tahsil where it covers only 8 per cent of net sown area. In the remaining tahsils of Tosham and Narnaul, wheat occupies about 17 and 26 per cent of net sown area respectively. This difference in percentage strength of wheat is the reflection of climatic and adaphic conditions.

Similarly, there are a lot of variations with reference to gross cropped area ranging from as low as 4 per cent in Siwani to as high as 34 per cent in Hisar. However, the percentage strength of wheat with respect to gross cropped area is almost in accordance with the net sown area.

7.2.3 Oilseeds

Oilseeds put together are third important crop of the region. It includes oilseeds of both the seasons, viz. *kharif* and *rabi*. It occupies 1,24,800 ha amounting to nearly 13 per cent of gross cropped area. There are many types of oilseeds, but the major ones are *toria*, *taramira*, *alsi* and *sarson* grown in *rabi* season, while ground nut, sesamum etc. are cultivated in *kharif* season. Weather plays decisive role in pollination of rape and mustard oilseed plants inflorescence affecting yield. Oilseeds plants specially *rabi* oilseed are very susceptible to adverse weather conditions. Frost and cloudy conditions are injurious to the crop. It requires moderate temperature and bright sunshine. Oilseed plants require loamy to sandy loam soils with adequate moisture. The plants require irrigation in case of low moisture in the soil.

The oilseeds in the subsistence economy of the region are next to foodgrains in its importance. Besides being the main source of edible oil, oilseeds also provide lubricant oil for making soap, varnish and paint, oilcakes to be used as nutrient food for cattle or fertilization of soil. Oilseed plants also increase soil fertility by fixing nitrogen and making more provision for aeration.

The distributional pattern of oil seeds is not uniform across the region (Table 7.5). It varies from as low as 7 per cent in Siwani to as high as 29 per cent of gross cropped area in Dadri. The data of oilseeds for Mahendragrah and Narmaul are not available and the researchers have also not been able to find it out. Therefore, the available data for remaining six tahsils have been presented in Table 7.5.

A very interesting picture emerges with respect to the percentage strength of oilseeds with reference to net sown area. It ranges from 14 per cent (lowest) in Siwani to 58 per cent (highest) in Dadri. In Bhiwani, Tosham and Loharu also it occupies 49.3, 43.8, and 47.2 per cent of net sown area respectively. In Hisar it covers only 16 per cent of net sown area. The highest coverage of net sown area by oilseeds is due to its cultivation in combination with gram, barley and such other crops.

Table 7.5: Spatial Pattern of Oilseeds (2005-06)

Tahsils	*Area in ha*	*Percentage to Total*	
		Net sown area	*Gross cropped area*
Hisar	19,700	15.89	9.61
Bhiwani	9,900	49.25	23.98
Tosham	26,600	43.82	21.29
Loharu	27,800	47.20	18.23
Siwani	6,200	14.06	7.12
Dadri	34,600	58.54	29.42
Mahendragarh	NA	—	—
Narnaul	NA	—	—
Total	**1,24,800**	**24.18**	**13.83**

Note: Data have been collected, streamlined and processed on the basis of village records by the author.

NA = Not available

The percentage coverage of oilseeds of gross cropped area also varies considerably, but it is almost in accordance with the percentage strength of net sown area. Leaving the tahsils of Hisar (9.61%) and Siwani (7.12%) it occupies more than one-fifth of the gross cropped area in the region. However, in view of the increasing demand of oil and its byproducts, not only the areal strength of oilseeds should be increased, but it should also be grown as a single crop.

7.2.4 Gram (Cicer Arietinum)

Gram is the major pulse crop of the region. It is best suited to area due to its low to moderate rainfall and a mild cold weather. Generally, it is cultivated as dry crop on light alluvial soil, well-drained sandy loam or loamy sand. Gram acts as an insurance crop requiring little care and as a useful catch crop on poor soils or in bad years. In case of dry soils, a light irrigation is sometimes given to it without flooding the field which is injurious for this crop. Gram increases the nitrogen content in the soil, and the well developed root system of the plant increases the aeration of the soil. It is very sensitive to alkaline soils. High water table areas are not suitable to its cultivation. Gram is most important winter food of the region.

Table 7.6: Spatial Pattern of Gram (2005-06)

Tahsils	*Area in ha*	*Percentage to total*	
		Net sown area	*Gross cropped area*
Hisar	6,800	5.49	3.31
Bhiwani	1,700	8.46	4.12
Tosham	22,400	36.90	17.94
Loharu	11,600	19.69	7.61
Siwani	33,400	75.74	38.57
Dadri	3,300	5.58	2.81
Mahendragarh	3,400	4.36	2.26
Narnaul	7,600	10.66	5.85
Total	**90,200**	**17.48**	**9.28**

Note: Data have been collected, streamlined and processed on the basis of village records by the author.

Gram is an important crop of the region and it occupies 90,200 ha amounting to 9.28 per cent of the gross cropped area. The spatial pattern of gram is shown in Table 7.6, which shows wide range of variation in its percentage strength, both with reference to net sown area and gross cropped area. It covers less than 10 per cent of the net sown area in 4 tahsils, viz. Hisar, Bhiwani, Dadri and Mahendragarh where more emphasis is given to the cultivation of *bajra* and wheat. But surprisingly, nearly three-fourths of net sown area in Siwani is devoted to its cultivation, while nearly 37 per cent in Tosham and 20 per cent in Loharu is given to gram cultivation. In the remaining tahsils almost 10 per cent or less than 10 per cent area is earmarked for gram cultivation. However, there is chequred pattern of its distribution.

The percentage strength of gram with references to gross cropped area is quite interesting. Except Siwani (38.57%) and

Tosham (17.94%) which form a contiguous belt in northern region leaving Hisar, in all the tahsils gram occupies less than 10 per cent of gross cropped area of the respective tahsils.

7.2.5 Other Pulses

The other pulses excluding gram, have a significant place in the region. They have their own importance in respect of their suitability for particular rotation or crop mixture or their adaptation to special agro-climatic conditions. Hence, they are of considerable regional and local significance. They enter largely into daily diet of the ruralites supplying the major portion of protein requirements of predominantly vegetarian peasantry of the region. Additionally, some of the pulses serve as excellent nutritious forage and grain concentrate to the large livestock population of the region. Some of the pulses are superior green manure crops adding the much needed humus and the plant nutrients to the soil.

There are many varieties of other pulses, such as, *urd*, black gram, *moong*, *moth*, brown gram, *masoor*, etc. All of them furnish useful leguminous rotation for increasing soil fertility. They constitute a group of crops of legume family which, with the help of the bacteria in their root module, have the power of fixing atmospheric nitrogen. Moreover, the well developed root system of pulses increases aeration of the soil. The spreading habits of some of the pulses provide a good cover to the soil against excessive evaporation and emission. Thus, they are evaporation reducing and erosion resisting crops either sown pure or in between the spaced rows of associated crops.

Pulses generally need less moisture than the companion crop and may survive when others are withering. Thus, the pulse crops are useful in improving the arable land by increasing the nitrogen content of the soil by drawing up the plant nutrient from the deeper layers, by serving or suitable green manure, and by acting erosion resisting and evaporation reducing crop. Environmentally too, sandy and excessively moist soils do not provide favourable adaphic adaptation for good yield of pulses. The soil and moisture requirement, vary from pulse to pulse.

However, other pulses put together occupy 91,200 ha covering about 9.38 per cent of gross cropped area. The spatial pattern of other pulses is given in Table 7.7.

Table 7.7: Spatial Pattern of Other Pulse (2005-06)

Tahsils	*Area in ha*	*Percentage to total*	
		Net sown area	*Gross cropped area*
Hisar	13,300	10.73	6.48
Bhiwani	1,700	8.46	4.12
Tosham	23,400	38.55	18.73
Loharu	1,300	2.21	0.82
Siwani	2,700	6.12	3.12
Dadri	3,800	6.43	3.23
Mahendragarh	NA	—	—
Narnaul	NA	—	—
Total	**91,200**	**17.67**	**9.38**

Note: Data have been collected, streamlined and processed on the basis of village records by the author.

The researcher has not been able to get the authentic data for the area under other pulses in Mahendragarh and Narnaul tahsils. The spatial pattern of other pulses is very much in accordance with adaphic and agro-climatic conditions. They occupy 91,200 ha amounting to nearly 18 per cent of net sown area or 9 per cent of gross cropped area. Except in Tosham tahsil where it covers nearly 38.55 per cent of net sown area, they cover almost 10 or less than 10 per cent of the net sown area. In Loharu they have covered just 2 per cent of net sown area, while it is 8 and 6 per cent in cases of Siwani, Dadri and Bhiwani tahsils respectively. The areal occupancy of other pulses in gross cropped area is similar to net sown area. Tosham has the highest percentage (18.73%), while lowest is recorded in Loharu. As these crops have many advantages over other crops, their areal strength should be increased.

7.2.6 Rice (Oriza Sativa)

Rice is generally confused with paddy which is a raw crop coming from the field. The rice grains enclosed in the hull is called paddy. It becomes rice only after the removal of hull. It requires a high atmospheric humidity, moderate temperature and well distributed rainfall between 1250 mm and 1500 mm during the growing season. It requires abundant supply of fresh water for irrigation when the rainfall is under 1250 mm. Rice thrives well on a variety of soils widely ranging from slightly acidic to slightly alkalic, particularly heavy compact and acidic soil, having the ability to hold water over the surface for a considerable period. The heavy alluvial soils are more favourable than a lighter type. There must be a sub-soil layer impervious to water underneath. Rice grown on clay and clayey loamy gives higher yield.

Due to arid and semi-arid conditions which lacks adequate soil moisture and less water resources, rice, being more water intensive crop, is not grown all over the region. The spatial pattern of rice is given in Table 7.8.

Table 7.8 reveals that rice is grown only in 3 tahsils, viz. Hisar, Bhiwani and Dadri. It covers 8,300 ha accounting nearly 1.61 and 0.93 per cent of the net sown area and gross cropped area respectively. It covers only 8.45 per cent of net sown area and 4.12 per cent of gross cropped area respectively in Bhiwani, while the corresponding percentages were 3.87 and 2.48, and 3.05 and 1.53 in case of Hisar and Dadri, respectively. Although, rice is a very important food grain of our country but its cultivation should

Table 7.8: Spatial Pattern of Rice (2005-06)

Tahsils	*Area in ha*	*Percentage to total*	
		Net sown area	*gross cropped area*
Hisar	4,800	3.87	2.48
Bhiwani	1,700	8.46	4.12
Tosham	—	—	—
Loharu	—	—	—
Siwani	—	—	—
Dadri	1,800	3.05	1.53
Mahendragarh	—	—	—
Narnaul	—	—	—
Total	**8,300**	**1.61**	**0.83**

Note: Data have been collected, streamlined and processed on the basis of village records by the author.

not be encouraged in this region which is deficient in water resources, because rice is a very water intensive crop. Therefore, its cultivation in the region will have adverse ecological implications.

7.2.7 Barley (Hordeum Vulgare)

Barley is a short duration drought scaping crop (Klages, 1957). During its short period of growth it requires moderate temperature and less supply of moisture. It can be grown successfully in many areas under unirrigated conditions. It is rarely irrigated except when it is sown intermixed with wheat because wheat needs more water either due to rain failure or improper conservation of soil moisture. Barley is a shallow root quick growing crop. It can be grown on all kinds of soils except the salt ridden water-logged and very light soil. Barley is a surface feeder, the roots are week and do not penetrate deep in the soil. It is preferably grown in combination with gram or oilseeds rather wheat and barley combination. Barley is a crop of rabi season sown in November and harvested in March/April. Barley is not a significant crop in the region. It covers 7000 ha. amounting to 1.36 per cent of net sown area or 0.72 per cent of gross cropped area. The spatial pattern of barley is given in Table 7.9:

Table 7.9: Spatial Pattern of Barley (2005-06)

Tahsils	*Area in ha*	*Percentage to total*	
		Net sown area	*Gross cropped area*
Hisar	3,300	2.66	1.61
Bhiwani	300	1.49	0.73
Tosham	400	0.65	0.32
Loharu	1,400	2.38	0.92
Siwani	100	0.23	0.12
Dadri	700	1.18	0.60
Mahendragarh	600	0.77	0.34
Narnaul	200	0.28	0.15
Total	**7,000**	**1.36**	**0.72**

Note: Data have been collected, streamlined and processed on the basis of village records by the author.

Table 7.9 shows that except two tahsils of Hisar (2.66%) and Loharu (2.38%), barley occupies one or even less than one per cent of the net sown area. Similarly, is the case with respect to its percentage share in gross cropped area in which, except Hisar (1.61%), it accounts less than one per cent in all tahsils. It shows the lower status of the crop. Its cultivation should not also be encouraged as it is neither remunerative nor soil conservative crop. Its nutritional value is also low.

7.2.8 Cotton (Gossypium Species)

Cotton is the foremost industrial and cash returning crop. Besides providing fibre, it also provides cotton seeds which furnishes a protein concentrate cake and a high grade oil. In fact, cotton cultivation in the region means life, health, happiness and prosperity (Singh, 1976). Climate is the determining factor for its cultivation.

It requires around 600 mm of rainfall with several days of bright warm weather between rains. Excessive soil moisture and high rate of transpiration are contributing factors determining the amount of shedding. At the time of picking, sun shine and absence of rain are preferable as the boll will burst with sufficient warmth. It requires a soil having good moisture holding capacity, favourable drainage and aeration. Such soils supplied with organic matter are most productive.

Cotton is very popular commercial crop of the region and it covers 71,600 hectares amounting to nearly 7 per cent of gross cropped area. The spatial pattern of cotton cultivation is given in Table 7.10.

Table 7.10: Spatial Pattern of Cotton (2005-06)

Tahsil	*Area in Ha*	*Percentage to total*	
		Net sown area	*Gross cropped area*
Hisar	46,400	37.43	22.62
Bhiwani	4,500	22.39	10.90
Tosham	7,000	11.53	5.61
Loharu	3,400	5.77	2.23
Siwani	2,900	6.58	3.49
Dadri	7,000	10.13	5.95
Mahendragarh	30	0.38	0.20
Narnaul	100	0.14	0.08
Total	**71,600**	**13.87**	**7.36**

Note: Data have been collected, streamlined and processed on the basis of village records by the author.

The spatial pattern of cotton cultivation is quite uneven ranging from as low as 0.14 per cent in Narnaul to as high as 37.34 per cent of net sown area in Hisar, but there is clear cut spatial pattern. Its areal strength decreases from north (the cotton belt) towards south. The zone of highest concentration is found in northern tahsils of Hisar, Bhiwani, Tosham and Dadri, where it occupies more than 10 per cent of the net sown area. Favourable climatic conditions, fertile sandy loam and loamy sand soil with humus content and adequate irrigation facility are the major causes of its concentration. In fact, it is the continuation of the main cotton belt of the state. The zone of moderate concentration with more than 5 per cent of net sown area is found in the middle of the region including Loharu and Siwani tahsils. The southern part of the region including the tahsils of Mahendragarh and Narnaul, where cotton occupies only 0.38 and 0.14 per cent of the net sown area, are unfit for cotton cultivation due to unfavourable climatic and adaphic conditions. The percentage coverage of cotton to gross cropped area is almost similar to net sown area but with a lower percentage. The largest coverage is in Hisar tahsil (22.62%) followed by Bhiwani (10.90%), while lowest is in Mahendragarh and Narnaul.

7.3 Crop-Combinational Analysis

Agricultural is, by far, an important single use of land in most of the countries of the world, and the description and analysis of various problems relating to agricultural activities have contributed a great deal to the content of the regional and economic geography of these countries. Such studies have, however, confined themselves to the description of dominant individual crops rather than complex agricultural system, With the result that they usually suffer from over generalizations because the individual crops, except in relatively rare circumstances of extreme monoculture, are not grown alone, but characteristically they appear in combinations (Weaver, 1954). Moreover, it is significant to note that any process of development has an inherited property of pushing the economic system from simpler to more complex. This complexity as manifested by diversification of places in terms of their economic activities can be measured by the different combinations of these activities.

Prior to the introduction of statistical technique, crop-combinations were indentified on the basis of the percentage strength of the crops. Monoculture were indentified on the basis of the first ranking crop, 2-crop-combinations on the basis of first and second ranking crops, 3-crop-combinations on the basis of first, second and third ranking crops and so on. But this intuitive classification based on subjective judgements like wheat zones, rice zones, industrial town, commercial town etc., have no statistical basis. All the places, in fact have multiple functions, but all of them may not be equally important at every place. Some of them may be very significant, while others might be quite inconsequential. A statistical basis is, therefore, necessary for identifying the group of significant functions, which represent the functional personality of an area (Mahmood, 1977).

Recently attention has been focused on combinational analysis of different elements in geographical studies which give a scientific treatment to the geographical problems. Geographers are particularly interested in classifying the places into different combinations of dominant activities including those which pertain to agricultural production and landuse. This method ultimately minimizes the chances of oversimplified generalizations too (Mohammad, 1975 and Athawale, 1996).

A general understanding of the particular combination of crops and the relative importance of each in an area can be very helpful in interpreting some aspects of social and economic geography of the region. J.C. Weaver has pointed out three different lines in which crop-combinational studies may have more significance. Firstly, crop-combination in itself is an integrated reality that demands definitions and spatial pattern analysis; secondly, crop-combination in itself is an integrated reality that demands definition and spatial pattern analysis; thirdly, such regions are essential for the construction of the still more complex structure of valid agricultural regions (Weaver, 1954).

Many methods have so far been devised by eminent scholars to identify and demarcate crop combination regions such as Quartile Method (Johnson, 1958), Mean Positive Deviation Method (Pownall, 1953), Standard Deviation Method (Nelson, 1955), Minimum Deviation Method (Weaver, 1954), Maximum Positive Deviation Method (Rafiullah, 1965), Modified Minimum Deviation Method (Doi, 1959), Successive Quotient Method (Kostrowicki, 1972), Maximum Distance Method (Ayyar, 1969) and so on. In this study Doi's Modified Minimum Deviation Method has been used because several crop combinational studies have found that in this method results are more realistic in comparison to other methods and they are not mere generalization. Moreover, the working with this method is much simple and easy.

7.3.1 Crop-Combination Regions

For identifying crop combination region Doi's method of Minimum Deviation has been used. Doi has modified Weaver's method of Minimum Deviation in two ways. Firstly, she has replaced Weaver's method $\frac{\sum d^2}{n}$ with only $\sum d^2$, and secondly, by working out a critical table for identifying crop-combinations. Doi has worked out a critical table in which for various possible crop combinations, the cumulative percentages along with the corresponding critical value are given in it. For working out the crop-combination in an enumeration unit, one has to work out the percentage of each crop occupying up to one per cent of the total cropped area of the unit and arrange them in descending order. For ascertaining monoculture, one has to locate the percentage of the first ranking crop in the percentage column of monoculture and compare the critical value given against this percentage in the table with percentage strength of the second crop. If the critical value is more than the percentage of the second crop, this crop is considered as insignificant crop and the unit is designated as monoculture area. But if the critical value of the table is less than the percentage of the second crop, this crop is considered as important crop and is included in the combination. Now to ascertain whether third crop is to be included in the combination or not, the percentage of first and second crops are added together and its position is located in the percentage column of 2-crop-combination and the critical value given against it is compared with the percentage of the third crop. If the critical value is more than the percentage of the third crop, this crop is considered as insignificant and the unit is identified as 2-crop-combination.

It can be illustrated by taking the example of a district where rice had occupied 42.46 per cent, wheat 35.87 per cent, sugarcane 10.63 per cent, oilseeds 3.60 per cent, pulses 1.40 per cent and maize 1.09 per cent. Since, first crop occupies 42.46 per cent and the critical value in monoculture column is zero in all cases of less than 51 per cent; the second crop is significant. Now, add the percentage of first and second crops (42.46 + 35.87 = 78.33), locate it in the column of 2-crop-combination and compare the critical value given against it, i.e. 20.0 with the percentage of third crop, i.e. 10.63. Since, the critical value is more than the third crop percentage, the crop is insignificant and the district is designated as 2-crop-combination region. Similarly, the position of subsequent crops is ascertained and crop-combinations are formed. Similarly, it can be illustrated by taking the example of on other district where wheat occupies 29.69 per cent, rice 22.49 per cent, *Bajra* 19.07 per cent, vegetables 4.84 per cent, pluses 3.60 per cent, barley 4.40 per cent and oil seeds 2.71 per cent.

Since the first crop occupies 29.69 per cent and the critical value is zero against all the percentages of less than 51 per cent. The second crop is significant. Add the percentages of first and second crops (29.69 + 22.49 = 52.18%) and locate it in the percentage column of 2-crop-combination and find out the critical value against it which is 1.03. Since, it is less than the percentage of third crop, this crop is significant. Now, add the percentage of first, second and third crops (29.69 + 22.49 + 19.07 = 71.25%) and located its position in the percentage column of 3-crop-combination which is 8.42. Since, it is more than the percentage of 4th crop, i.e. 4.84, this crop is insignificant and the area is designated as 3-crop-combination.

As a result of statistical processing of data based on Doi's method using Critical Table, the crop-combinations in each tahsil have been worked out. The results are given in Table 7.11.

Table 7.11: Crop-Combination Region (2005-06)

Crop-Combination	*Crops*	*Tahsil*
2 Crop-combination Regions	*Bajra*-Oilseeds	Dadri
	Bajra-Gram	Siwani
	Bajra-Wheat	Mahendragarh
3 Crop-combination Regions	Wheat-Cotton-*Bajra*	Hisar
	Bajra-Oilseeds-Wheat	Loharu
	Bajra-Wheat-Gram	Narnaul
4 Crop-Combination Regions	Oilseeds-*Bajra*-Wheat-Cotton	Bhiwani
	Bajra-Oilseeds-Pulses-Gram	Tosham

Note: Data have been collected, streamlined and processed on the basis of village records by the author.

It is clear from Table 7.11 that monoculture is non-existent in the region. It is interesting that 2, 3 and 4 crop-combinations have emerged, but none of the tahsils have the same crops in the same sequence. *Bajra* is the only crop which has come in combination

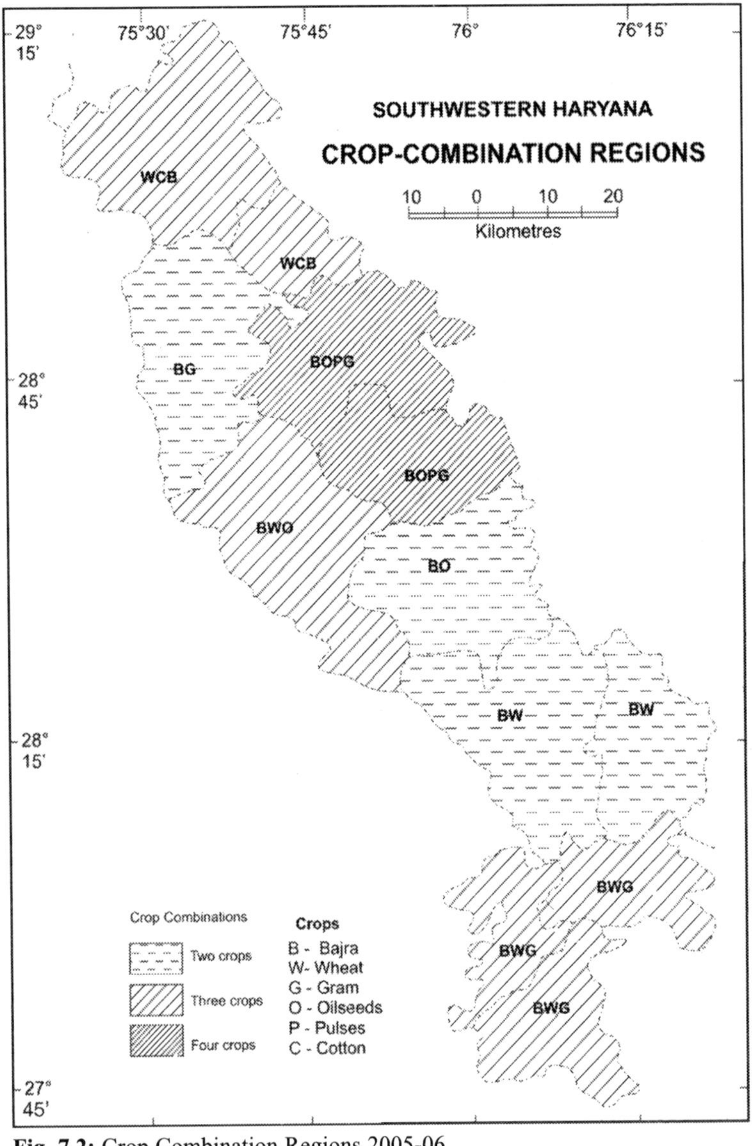

Fig. 7.2: Crop Combination Regions 2005-06

of all the tahsils. All together 6 crops viz. *bajra*, oilseeds, gram, wheat, cotton, pulses have emerged in different combinations and sequences. A brief description of each crop-combination region would be worthwhile.

7.3.1.1-2 Crop-Combination Region

There are three tahsils where 2-crop-combination has emerged, and *Bajra* is the first crop in all of them. In Dadri *Bajra*-Oilseed combination, in Siwani: *Bajra*-Gram combination, and in Mahendragarh, *Bajra*-Wheat combination have emerged. It shows that *bajra* dominates the agricultural scene in *kharif* season, but in *rabi* season a single crop has not emerged as important crop in all tahsils. However, Dadri and Mahendragarh have formed contiguous region.

7.3.1.2-3 Crop-Combination Region

There are also three tahsils where 3 crop-combination has emerged. Here 5-crops have come in combination. *Bajra* and wheat have come in combination of all tahsils, but in different sequences. However, in Hisar Wheat-Cotton-*Bajra*, in Loharu. *Bajra*-Oilseeds–Wheat, and in Narnaul *Bajra*-Wheat-Gram have come in combination. None of these combination regions form a contiguous region.

7.3.1.3-4 Crop-Combination Region

Four crop-combinations have been found in two tahsils, i.e. Bhiwani and Tosham with Oilseeds-Bajra-Wheat-Cotton, and Bajra-Oilseeds-Pulses-Gram crop-combinations respectively. The above analysis shows that there is not much agricultural diversification in the region.

7.4 Levels of Agricultural Development

As a result of 60 years of planned development programme and policies, overall improvement in socio-economic conditions has taken place. The structure of national and state economics has changed. The new technology in agricultural production and industry and the socio-institutional framework within which economic activities are conducted have substantially under gone change. The socio-economic condition of the masses has considerably improved. The literacy level, housing condition, quality of life have gone up. But the level of development has not been uniform at any level. Inter and intra-sectoral differences in economic structure become more sharp and noticeable. So is the case with intra and inter-regional level of development. As such the regional disparities in level of agricultural development also kept on aggravating during these plan periods all over the region. Therefore, the level of agricultural development has been assessed for the year 2005-06.

For identification, classification, regionalization, and even for exploring casual relationship in explanatory model as many as 21 indicators, the value of which differ from tahsil to tahsil, have been selected which have been grouped into following four sub-categories:

Categories	*No. of indicators*
A. Agricultural Condition	6
B. Agricultural Infrastructure-I	5
C. Agricultural Infrastructure-II	6
D. Agricultural Production	4

In selection of large number of indicators, mainly two basic problems arise, viz. problem of co-linearity and conversion of these indicators into comparable unit and contracting them into few indices in order to compare inter tahsil level of development. For the former Correlations Matrices have been worked out and the indicators having very high level of correlation have been deleted. The latter has been a matter of research of various discipline for the last 5 decades. However, the major methods developed for this purpose are preparation of composite index with the help of scores, (Mohammad, 1992), ranking methods (Kendall, 1939) and principal component analysis/factor analysis (Singh, 1980). All these methods have their own merits and limitations. Large number of studies have been done to workout composite index of development either by simple aggregation of ranking of areal units or by assigning weights to selected indicates. In the present study simple aggregation of ranking of areal units has been followed. Firstly, all the indicators have been converted into comparable unit by standardizing them with the help of following formula which is known as Z value or Z score.

$$Z_{ij} = \frac{x_{ij} - x_j}{\delta x_j}$$

where

Z_{ij} = standardized value of the indicator;

X_{ij} = Value of X_j Variable in the observation;

X_i = Mean value of x_i variable; and

δX_j = Standard deviation of X_j variable.

After working out the Z score of all indicators, average is taken out for each group of variables in all the 4 sub-categories. Since, all the 4 categories are not equally significant nor they equally contribute to the agricultural development of the region, weightage has been given to each of them as given below:

Agricultural Condition	0.3
Agricultural Infrastructure-I	0.2
Agricultural Infrastructure-II	0.1
Agricultural Production	0.4
Total	**= 1.0**

In order to assess overall development, the sub-category averages have been multiplied with its respective given weightage. The resulting figures of all these averages have been added together and divided by 4 which gives the composite index of agricultural development which is given in Table 7.12.

Table 7.12: Composite Index of Agricultural Development (2005-06)

Categories	*Score range*	*Tahsils*
Medium	>0.50	Hisar-Bhiwani-Tosham
Low	0.50-0.00	Dadri-Mahendragarh
Very Low	0.00-0.50	Siwani-Loharu-Narnaul
Total		

Note: Data have been collected, streamlined and processed on the basis of village records by the author.

On the basis of this result, the area has been divided into following regions of agricultural development:

7.4.1 The Region of Medium Level of Development
7.4.2 The Region of Low Level of Development
7.4.1 The Region of Very Low Level of Development

7.4.1 The Region of Medium Level of Development

Table 7.12 shows that three tahsils, *viz.* Hisar, Bhiwani and Tosham have medium level of development with a composite index of more

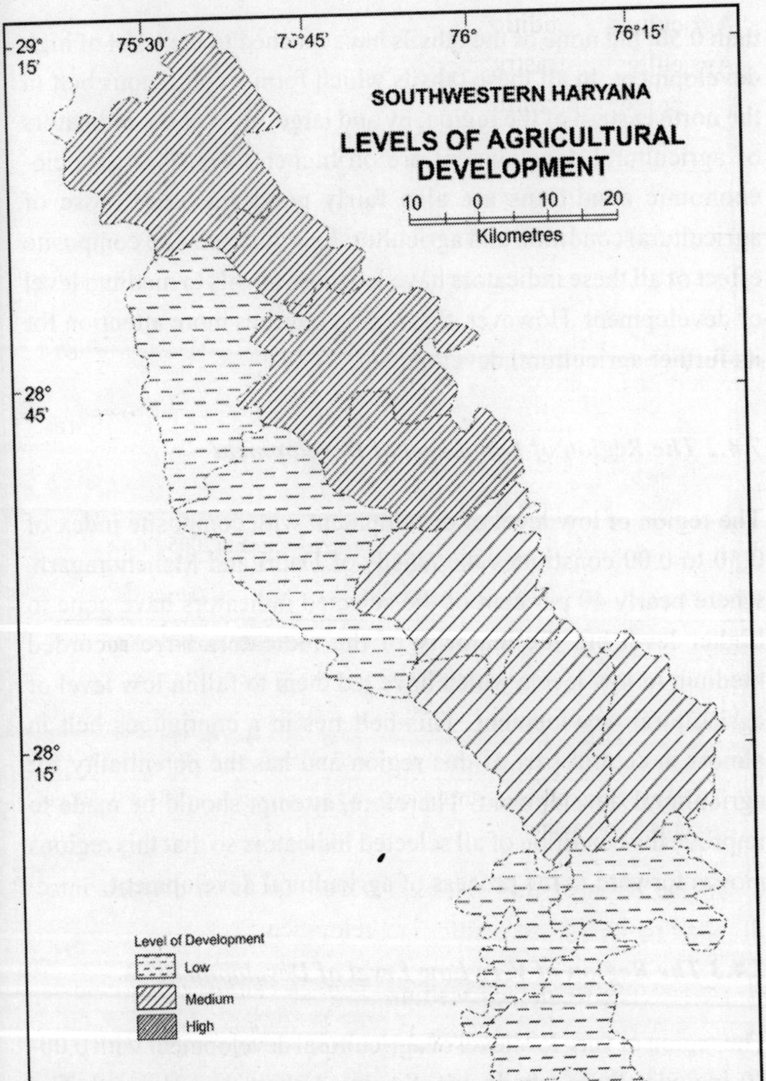

Fig. 7.3: Levels of Agricultural Development in South-western Haryana

than 0.50, but none of the tahsils have reached to the level of high development. In all these tahsils which form a contiguous belt in the north eastern of the region, by and large, most of the indicators of agricultural development are on higher sides. Physio-socio-economic conditions are also fairly good specially those of agricultural condition and agricultural production. The composite effect of all these indicators have led these tahsils to medium level of development. However, this region requires more attention for its further agricultural development.

7.4.2 The Region of Low Level of Development

The region of low level of development with composite index of 0.50 to 0.00 constitutes the tahsils of Dadri and Mahendragarh, where nearly 40 per cent of the selected indicators have gone to higher level but the majority of the indicators have recorded medium to low levels which have led them to fall in low level of agricultural development. This belt lies in a contiguous belt in almost in central part of this region and has the potentiality for agricultural development. Therefore, attempt should be made to improve the condition of all selected indicators so that this regions moves forward in the process of agricultural development.

7.4.3 The Region of Very Low Level of Development

The region of very low level of agricultural development with 0.00-50 includes three tahsils, *viz.* Siwani, Loharu and Narnaul. The former two tahsils lying in western part of the region constitute a compact belt, while Narnaul lies in the extreme southern part of the region. In this region the increasing level of arid condition in western part on the one hand and the continuation of Aravali range with its out crops and rock exposure on the other hand are the main causes of very low level of agricultural development. This region has little scope for further improving the conditions of agricultural development indicator because it may adversely affect the ecological condition of the region.

References

Ahmad, M.S., 2004: *Five Decades of Planning and Developing: A Case Study of Uttarakhand.*

Athewala, A.G., 1966: Some New Methods of Crop Combination, *Geographical Review of India,* Vol. XXXVIII, No. 1, Kolkata.

Ayyar, N.P., 1969: Crop Regions of Madhya Pradesh, *Geographical Review India*, Vol. XXXI, No. 1, Kolkata.

Bader, M.D.M., 2010: *Disparities in Neighbourhood for Food Environment: Implications, Measurement and Strategies.*

Borlaug, N.E., 2004: Feeding a World of 10 Billion People: Our 21st Century Challenge, in Scaners, C.G. and Milinowski, I.A. (eds.), *Perspectives in World Food and Agriculture,* Blackwell, Two.

Chikara, D.V., 2006: *Levels of Agricultural Development in Haryana* (unpublished Ph.D. Thesis) Department of Geology, Delhi University, Delhi.

Doi, K., 1959: The Industrial Structure of Japanese Frefacture, *Proceeding IGU Regional Conference,* Japan.

Government of India: *Agricultural Statistics at a Glance*, 2006 to 2012.

Government of India, *Indian Agricultural Statistics*, Vols. I & II, New Delhi.

Hooda, R.S., 1994: Impact of Sprinkler Irrigation on Crop Combination: A Case Study of Bhiwani District, Haryana, *Crop Research*, Vol. VIII, No. 1.

Mahmood, A., 1977: *Statistical Methods in Geographical Studies*, Rajesh Pub, New Delhi.

Majeed, A., 1992: Crop-combinational Analysis: A Review in Methodology, in Mohammad, N. (ed.) *New Dimensions in Agricultural Geography*, Vol. 5, Concept Publishing Co., New Delhi.

Mohammad, N., 1982: 'Levels of Agricultural Development in U.P.' in Mohammad, N. (ed.) *Perspective in Agricultural Geography*, Vol. 4 Concept Publishing Co., New Delhi.

Mohammad, N. and Arora, M., 2011: Role of Urban Agriculture in Food Security in North-western Delhi, in Rana, N.K. and Singh, K.N. (eds.) *Holistic Development: A Geographical Perspective*, R.K. Book, New Delhi.

Mohammad, N., Munir, A., and Rehman, A., 2007: *Fifty Years of Indian Agriculture*, Concept Publishing Co., New Delhi.

Mohammad, N. and Raza, M., 1992: "Agricultural Meterological Data Base and Agricultural Productivity" in Mohammad, N. (ed.) *New Dimension in Agricultural Geography*, Concept Publishing Co., New Delhi.

Mohammad, N., 1975: Crop Association Regions in Ghaghara Rapti Doab, *The North Eastern Geographer*, Vol. VII.

Mohammad, N., 1978: *Agricultural Landuse in India*, Inter-India Pub., New Delhi.

Nag, P., 2010: *Atlas of Agricultural Resources of India,* NATMO, Kolkata.

Patra, P., 2006: *Land Degradation in Tropical Highlands*, Rajesh Pub., New Delhi.

Rafiullah, S.M., 1965: A New Approach to Functional Classification of Towns, *The Geographer*, Vol. 22, Aligarh.

Singh, J. and Dhillon, S.S., 1984: *Agricultural Geography*, Tata McGraw Hill, New Delhi.

Singh, R.P., 2007: 'The Role of Bio-fertilizer in Crop Production' in Sharma, A. and Singh, R.K., (eds.) *Composite Farming Practice and Economic Development*, Abhijeet Pub., New Delhi.

Sinha, R.R.K., 2009: *Dynamic of Land Caste Relationship,* Manak Pub, New Delhi.

Vaidyanathan, A., 2010: *Agricultural Growth in India: Role of Technology Incentive and Institution*, Oxford University Press, New Delhi.

Weaver, J.C., 1954: Crop-Combination Regions in Middle West, *Geographical Review*, Vol. XLIV, No. 2.

8
Sprinkler Irrigation and Socio-Economic Transformation

A Micro-Level Analysis

8.1 Introduction

Agriculture is not only the product of physical environment, but also a set of techno-economic and socio-cultural framework such as adoption of agricultural innovations, land ownership, size, shape and layout of farms, educational level, age structure, agricultural implements, market centres etc. There is no general theory of socio-eco-cultural composition which could be fairly applied to explain the spatial variations in agricultural system, but social and economic differences within communities and between communities do affect the scale and type of farming operations together with choice of enterprises. The decision of the farmers to put the land to different uses depends upon their age, education, sex, occupation, psychology and so on (Bose, 1961; Lee, 1960 and Mohammad, 1988). In this chapter, therefore, an attempt has been made to discuss these and other personal, household and farming characteristics of the respondents to assess the status of adoption of sprinkler irrigation, level of agricultural development and socio-economic transformation.

8.2 Personal and Household Characteristics

The personal and household characteristics of the respondents and their families have a direct bearing on primary occupation specially agriculture, exposure to impersonal communication, adoption of agricultural innovations, level of knowledge and involvement etc. Therefore, some of the important socio-cultural components have been analysed here. Here, an attempt has been made to find out and explain the differences in all these parameters among user and non-user respondents of sprinkler irrigation.

8.2.1 Age Composition of Respondents

Adoption of new idea, practices and technology helps to develop the agricultural system of an area. In this process age composition plays an important role. The presumption is that at a comparatively younger age people are more receptive to new ideas and practices, whereas at the old age they find it difficult to change from age old practices and they resist to adoption of innovations (Mohammad, 1988). With this view all the 465 respondents of the 13 selected villages have been arranged in ascending order and have been grouped into 4 categories according to age groups (Table 8.1).

Table 8.1: Age Composition of Respondents

Age Group		*No. of respondents*		*Total*
		Users Res.	*Non-users Res.*	
Young Age	(<30 yrs)	132 (40.00%)	18 (13.34%)	150 (32.26%)
Middle Age	(31-45 yrs)	105 (31.82%)	23 (17.02%)	128 (27.53%)
Upper Middle Age	(46-59 yrs)	67 (20.30%)	38 (28.15%)	105 (22.58%)
Old Age	(>60 yrs)	26 (7.88%)	56 (41.49%)	82 (17.63%)
TOTAL		**330 (100%)**	**135 (100%)**	**465 (100%)**

Source: Collected, compiled and computed by the author.

Table 8.1 shows that there are 150 respondents accounting a little less than one-third of the total respondents are in the young age-group of less than 30 years. The largest number of respondents, i.e. 233 amounting to nearly half of the total are in the middle-age group between 31 and 59 years. The old age respondents of more than 60 years account nearly 18 per cent of the total respondents, while the upper middle age group respondents are 105 account nearly 22.58 per cent of the total respondents.

The above analysis clearly indicates that the younger population, who is comparatively more educated and innovative, either takes little interest in agriculture and is interested in white collar jobs or is being given little freedom to agricultural operations. The middle age respondents within the age-group of 31 to 60 amounting to 50 per cent of the respondents take more interest in agricultural activities. Interestingly nearly 18 per cent of respondents belong to old age group take much interest in agricultural activities. It shows that social values and norms dominate the agricultural activities. By and large, father or elder brother, by virtue of their being older in age, is not only the leader of the family, but his decision is final in day-to-day routines including agricultural activities.

Table 8.1 further shows that there are a lot of differences in the age-group of users and non-users of sprinkler irrigation. The users account 40 per cent of its total belonging to the young age. The lowest percentage belongs to old age. It means that with increase in age the percentage declines, while in case of non-users opposite is the case. It is also clear from Table 8.1 that the user respondents of sprinkler irrigation, by and large, are younger in age as compared to their counterpart.

8.2.2 Educational Status of Respondents

The role of education in the development of agriculture hardly needs any emphasis. It is one of the chief instruments through which a society socializes its members and brings desirable changes in the social life of its people. In fact education is the aggregate of all the processes by means of which a person develops abilities, attitudes, and other forms of behaviour of positive values in the society he lives or it is a social process by which people are subjected to the influence of selected and controlled environment, so that, they can attain social competence and optimum individual development (Bear, 1947). And, as such it could be argued that education influences the course of behaviour of both society and its individuals. And, adoption of farming occupation and related activities are not exception. Various studies have proved that there is a close relationship between educational level and adoption of agricultural innovations which in turn affects agricultural diversification (Mohammad, 2013; and Rangaswamy, 1972).

However, in order to ascertain the educational status of the individual respondents, questions on formal and informal education of respondents were included in the questionnaire. Different answers varying from being illiterate to Ph.D. were received.

Literacy Standard	*No. of Scores*
Illiterate	0.00
Informal Education	0.25
Primary School	0.50
Middle School	0.75
High School	1.25
Senior Secondary	1.75
Graduation	2.50
M.A. & Others	5.00

So, it was difficult to classify the respondents into any meaningful category on this basis. Therefore, in order to assess the educational status of respondents at uniform level, and also to classify them into reasonable categories, following scoring scheme has been adopted (Verma, 2007). This scheme may be subjective, but utmost care has been taken to maintain objectivity.

After assigning the scores, the total scores of every individual was found out which were arranged in ascending order. Finally, the respondents were divided in 4 categories as given in Table 8.2.

Table 8.2: Educational Status of Respondents

Groups	*Score range*	*No. of Respondents*		*Total*
		Users Resp.	*Non-user Resp.*	
Illiterate	0.00	27 (8.17%)	13 (9.63%)	40 (8.6%
Less Educated	<1.50	63 (19.10%)	28 (20.74%)	91 (19.57%
Moderately Educated	1.50-2.50	48 (44.85%)	41 (30.37%)	189 (41.65%
Highly Educated	> 2.50	92 (27.88%)	53 (39.26%)	145 (31.18%
Total		**230 (100%)**	**135 (100%)**	**465 (100%**

Source: Collected, compiled and computed by the author.

Table 8.2 indicates that nearly 8.6 per cent of the respondents are illiterate, while one-fifth of the respondents are less educated with less than 1.50 educational score. There are 189 respondents accounting about 41 per cent of the total are moderately educated, while nearly 31 per cent respondents belong to highly educated category with more than 2.5 educational score. It may be inferred from the above tabular analysis that the literacy level among the respondents is moderate. The picture in case of educational status of users and non-users of sprinkler irrigation is opposite to each other. By and large, the users are more literate than non-users.

8.2.3 Educational Index of the Household

Education plays an important role in the decision-making process. In the overall process of agricultural development and agricultural diversification, it is not only the educational status of the respondents that matters but also the educational level of all the family members. Educated households easily accept innovations in every field as compared to less educated households. The surveyed household members have educational status ranging from informal education to post-graduation. Hence, it is not possible to aggregate the educational status of the household due to varied size of family and different educational status. Therefore, the educational index of the household has been worked out with the help of scoring scheme given earlier. After assigning scores to the family members, the scores were added together and divided by total members of the household excluding children below 5 years of age. The results thus obtained have been arranged in ascending order and the households have been divided in 5 categories (Table 8.3).

Table 8.3 indicates that none of the households are under illiterate category, while nearly 28 per cent households fall under

Table 8.3: Educational Index of Household

Groups	*Score range*	*No. of Respondents*		*Total*
		Users Resp.	*Non-users Resp.*	
Illiterate	0.00	—	—	—
Less Educated	<1.50	104 (31.52%)	25 (18.52%)	129 (27.74%)
Moderately Educated	1.50-2.50	142 (43.03%)	73 (54.07%)	215 46.24%)
Highly Educated	>2.50	84 (25.45%)	37 (27.41%)	121 (26.02%)
Total	—	**330 (100.00)**	**135 (100.00)**	**465 (100.00)**

Source: Collected, compiled and computed by the author.

less educated category. There are 215 respondents accounting nearly 46 per cent of the total are moderately educated, while 26 per cent respondents are highly educated. But the educational index of users and non-users households is dissimilar.

A close look at the household educational level of the households clearly indicates that in the villages located in the vicinity of plains or major towns, the educational level is high due to availability of educational infrastructure, while in the remaining areas the educational level is low. Table 8.3 further shows that the educational index of the users household is higher as compared with the non-users because they cannot only afford the education of their family members but are also more conscious about it.

8.2.4 Family Size of Respondents

The old agrarian societies of the orient are characterized as familists, and the whole social organization and the values and beliefs of the people is centred around social institutions. Most of the families, specially in the countryside, are large in size and extended or joint in composition. Parents generally live with their married sons and their families. By and large, it is the family, instead of the individuals, which is the unit of social action. The thought and actions of the individual members in the family, in a large measure, are influenced by the family as a single entity (Mohammad, 1988). The personality of a member of a large family is somewhat different from that of a member belonging to a small family. Hence, the size of a family has a major role in deciding the type of occupation to be adopted as well as in the process of its development.

Therefore, the total members of the selected respondents were enumerated during the course of the fieldwork. In the present study

Table 8.4: Size of Family of Respondents

Family Size	*Members range*	*No. of Respondents*		*Total*
		Users Resp.	*Non-users Resp.*	
Small	<4	113 (34.24%)	27 (20.00%)	140 (30.12%)
Medium	5-9	97 (29.39%)	55 (40.74%)	152 (32.69%)
Large	10-14	61 (18.48%)	28 (20.74%)	89 (19.14%)
Very Large	>14	59 (17.89%)	25 (18.52%)	84 (18.05%)
Total		**330 (100%)**	**135 (100%)**	**465 (100%)**

Source: Collected, compiled and computed by the author.

the family includes all the members of selected respondents who live in one household and share a common kitchen. On the basis of the total members of the households, households were arranged in ascending order and finally, grouped into five categories as given in Table 8.4 which shows that on the whole the size of the family of respondents is reasonable. There are 140 respondents amounting to nearly 30 of the total respondents who have small size of family with less than 4 persons, while 152 respondents amounting to 33 per cent of the total respondents have a medium size of family with 4 to 9 members. Most of these respondents belong to the north-western region of the state where the means of sustenance is poor and they are leading a harsh life. Similarly, nearly one-fifth of the respondents have a large size of family, while only 84 respondents accounting 18 per cent of the respondents have very large size than 10 members in it.

Table 8.4 further shows that, by and large, the users of sprinkler irrigation have comparatively small size of family, e.g. 34 per cent user respondents have small size of family, while in case of non-users it is 20 per cent. Similarly, 29 per cent users have medium size of family, while non-users have nearly 21 per cent under this category.

8.2.5 Attitude Towards Family Welfare Programme

The explosive rate of growth of population is considered to be the greatest menace in process of development of our country. And, therefore, various measures which have not been uniformly adopted by the people of all castes, have been suggested to curb the population growth. Earlier the rural masses, specially the weaker section had conservative approach regarding the size of family. They considered large size family as a measure for social security and instrument for generating economic pursuits. According to them children are the gift of nature and any measure to prevent it is really a sin. But the younger and educated section of the society are changing their structure and system of the family. Since, it is linked with religion and economic gain, it will take comparatively longer time to change, but the process has been set in motion. However, for the assessment of the attitude of the respondents towards family planning programme a number of questions related to awareness, mode, attitude and adoption have been included in the questionnaire. The family planning programme, its adaptation and impact are very sensitive cultural issues. Therefore, in the formulation of the questionnaire utmost care has been taken so that it does not hurt their feelings, norms and values. Since, the nature of the questions and their response was of different nature, it was not possible to aggregate the data. Hence, weighted scores have been given to responses of the respondents, and accordingly they have been grouped into three categories (Table 8.5).

It is evident from Table 8.5 that 41 per cent of the respondents are at a lower web in case of adoption of family planning programmes. It shows their strong bonds with rituals, traditions, customs and religion. However, there is hope of adoption of this programme because a little less than one-third respondents belong to medium category, while about 19 per cent are in the higher

Table 8.5: Index of Family Planning of Respondents

Category	*Score Range*	*No. of Respondents*		*Total*
		User resp.	*Non-user resp.*	
High	More than 3.50	137 (41.52%)	51 (37.78%)	188 (41.43%)
Medium	3.50-2.50	101 (30.60%)	37 (27.41%)	138 (29.68%)
Low	Less than 2.50	92 (27.88%)	47 (34.81%)	139 (18.89%)
Total		**330 (100%)**	**135 (100%)**	**465 (100%)**

Source: Collected, compiled and computed by the Author.

category. Table 8.5 further shows that there is difference between the user and non-user of sprinkler irrigation. Generally, the users have shown more interest and adoption as compared with their counterpart.

8.2.6 Political Awareness and Involvement

The politics is so deeply interwoven in the fabrics of our daily life that no facet of it can remain in political isolation. The political behaviour of Indian masses can no more be taken for granted, but they have to be wooed, conversed and cajoled. They have gradually learnt the art of claiming their share of national cake. The democratic process involving different elections right from the Member of Parliament down to the *Panchayat* elections has created awakening, interest and involvement among the rural masses. The scheduled castes and the Muslim minority who have a major consolidated vote bank have also become more conscious about their political rights and privileges. But their level of awakening is not uniform across the board. There are a number of factors which contribute to it. Therefore, an attempt has been made to assess the political awareness and involvement among the people by imposing a number of questions related to national and international political affairs. Queries were made regarding their involvement at national level and local level politics. Certain questions related to their political perceptions were also posed to the respondents. As the nature of the questions posed and the responses received were not of the similar nature and character, it was difficult to aggregate them. Therefore, the responses of the respondents were given weighted score which have been used to work out total score of individuals. Based on it, the respondents have been grouped into various categories.

Table 8.6 showing the degree of political awareness and involvement shows that the respondents are uniformly distributed in all the categories amounting to roughly one-third in each one of them. Almost the similar pattern is observed in case of user and non-user respondents.

Table 8.6: Level of Political Awareness and Involvement of Respondents

Category	*Score Range*	*No. of Respondents*		*Total*
		User resp.	*Non-user resp.*	
Less Involved	<1.00	107 (32.42%)	48 (35.36%)	155 (33.34%)
Moderately Involved	1.00-1.50	101 (30.62%)	49 (36.30%)	150 (32.26%)
More Involved	>1.50	122 (36.96%)	38 (28.14%)	160 (34.40%)
Total		**330 (100%)**	**135 (100%)**	**465 (100%)**

Source: Collected, compiled and computed by the Author.

This statistical analysis clearly indicates the higher involvement of the masses specially the scheduled castes into the national and local level politics.

8.2.7 Level of Social Transformation

Ever since Independence, India has formulated and implemented various programmes and policies through its various Five Year Plans for the social upliftment of the masses in general and scheduled castes in particular. The Government has introduced the programme of free primary education, scholarship to scheduled castes and backward castes, mid-day meal, free distribution of foods grains, sale of foodgrains at much lower prices to weaker section, family welfare programme, various incentives to adoption of family planning programmes and many other similar programmes from time to time for the social upliftment of the weaker section of the society. As a result there have been considerable changes in various social parameters of the people in general and the scheduled castes in particular. Among various social parameters, age structure, educational levels, attitude towards marriage, family size, attitude towards family welfare programme, political awareness and involvement, and social mobility have been considered here

In the proceeding discussion the social transformation has been described and explained with respect to 6 selected indicators. All of them have shown positive and encouraging level of change at differential levels. As these indicators are of different nature and character and have been given different scoring schemes, they could be directly averaged out in order to get overall level of social transformation. Therefore, a composite Index of social transformation with the help of the following formula has been worked out:

$$CIT = \frac{\sum_{x=1}^{n} x_{ij}}{n}$$

where

CIT = Composite index of transformation;

x_{ij} = score value of ith indicator; and

n = Number of indicators.

The result thus obtained is 3.10 in which 0.40 is added and a rounded scale of more than 3.50 scores (High), 3.50-2.50 scores (Medium) and less than 2.50 scores (Low) has been taken to classify the respondents. Consequently, the respondents have been grouped into three categories based on their individual scores of social transformation (Table 8.7).

Table 8.7 gives very encouraging result of social transformation indicating that about 40 per cent of the total respondents have high level of social transformation, while more than 32 per cent fall in the medium category. The scrutiny of individual score of transformation of respondents indicates that about half of respondents have scores higher than the standard scale of social transformation.

Table 8.7: Composite Index of Social Transformation of Respondents

Category	*Score Range*	*No. of Respondents*		*Total*
		User resp.	*Non-user resp.*	
High	> 3.50	154 (46.67%)	34 (25.19%)	188 (40.43%)
Medium	3.50-2.50	107 (32.42%)	43 (31.85%)	150 (32.26%)
Low	< 2.50	69 (20.91%)	58 (42.96%)	127 (27.31%)
Total		**330 (100%)**	**135 (100%)**	**465 (100%)**

Source: Collected, compiled and computed by the Author.

However, 127 respondents amounting to nearly 27 per cent of the respondents have low level of transformation. Table 8.7 further shows that on the whole the users of sprinkler irrigation have higher level of social transformation and their number is also more. This proves our proposed hypothesis, viz. "Sprinkler irrigation leads to agricultural development which in turn leads to social transformation."

8.3 Level of Economic Transformation

Ever since Independence, Government has formulated and implemented various programmes and policies for the overall development of the country. Many of these programmes had been directly related with the upliftment of the rural masses. As a result their economic condition has considerably improved. Hence, an attempt has been made to assess and explain the economic transformation of the selected respondents.

For the economic transformation there could be several indicators, but it is very difficult to generate data for all these indicators. Therefore, for the assessment of economic transformation income level, standard of living, housing condition, exposure to mass media and operational size of holding, have been taken into consideration. A brief account of each of these is given here.

8.3.1 Level of Per Capita Income

After Independence, India has been persistently trying to increase the per capita income so as to raise the standard of living of masses by inducting various programmes and policies during different Five Year Plans which have actually started bearing the fruit after mid-sixties when the Green Revolution Programme was launched. Subsequently, various other programmes of rural upliftment especially for weaker and downtrodden section of the society were introduced which have diversified the occupational structure of the rural masses and subsequently have helped in increasing their income level. They have switched over to more remunerative occupations. Consequently, per capita income of scheduled caste population has also increased. Although per capita income is very good indicator for the assessment of economic status, but the collection of data especially in rural set-up of diversified occupational structure with no fixed income or daily wages is a Herculean task. It is more so when there is no system of maintaining the record of the income. However, in order to work out per capita income of the respondents, the earnings of all family members in whatsoever form it was, were ascertained by putting several questions in different forms. The total income of the household, thus ascertained, was converted into monetary value with a uniform standard conversion scale, and finally divided with the total members of the household. The results thus obtained gives per capita per annum income of the family member which has been compared with their earlier calculated per capita per annum income. Subsequently, weighted scores have been given and level of change is ascertained. Finally, based on the results, the respondents have been grouped into three categories (Table 8.8).

Table 8.8: Index of Income Level of Respondents

Category	*Income Range*	*No. of Respondents*		*Total*
		User resp.	*Non-user resp.*	
High	> 15,000	137 (41.52%)	38 (28.15%)	175 (37.63%)
Medium	15,000-12,000	109 (33.03%)	29 (21.48%)	138 (29.68%)
Low	< 12,000	84 (25.45%)	68 (50.37%)	152 (32.69%)
Total		**330 (100%)**	**135 (100%)**	**465 (100%)**

Source: Collected, compiled and computed by the Author.

Table 8.8 shows that around 37 per cent of the total respondents fall under low-income group, while about 30 per cent fall under medium category. As such nearly 62 per cent respondents do not have enough income to maintain a reasonable standard of living. About 38 per cent respondents belong to higher income group. Table 8.8 further shows that the income level of user respondent is higher than non-user respondent. Nearly 42 per cent of the user respondents have higher income, while only 28 per cent of the non-users fall under this category. Nearly half of the non-user respondents have low income while only one-fourth of the user respondents are in this category. This is mainly due to use of sprinklers which have increased their agricultural production and consequently their income.

8.3.2 Level of Standard of Living

In fact the standard of living is a very effective indicator to gauge the socio-economic condition of the people. Although mere observation and making qualitative comments of standard of living is easier, but its quantitative measurement is comparatively difficult because it is, by and large, manifestation of income, housing condition, dress and dressing materials, household utensils, exposure to mass media, social status and a host of such conditions. Per capita income is, perhaps, the main factor, but it is not everything for the assessment of standard of living. However, in order to gauge it, various pin pointed and inferential questions were framed and administered on various possible parameters considered to be contributing to standard of living. As the nature of quarries and the responses of the respondents were of varied nature, judicious scores have been given to them, and on this basis the standard of living and the changes therein have been worked out for all the respondents who have been classified into three categories according to the scale of standard score of change (Table 8.9).

Table 8.9: Index of Standard of Living of Respondents

Category	*Score Range*	*No. of Respondents*		*Total*
		User resp.	*Non-user resp.*	
High	>3.5	156 (47.27%)	29 (21.48%)	185 (39.78%)
Medium	3.5-2.5	103 (31.21%)	43 (31.85%)	146 (31.40%)
Low	<2.5	71 (21.52%)	63 (46.67%)	134 (28.82%)
	Total	**330 (100%)**	**135 (100%)**	**465 (100%)**

Source: Collected, compiled and computed by the Author.

The above analysis and observation of the collected facts and semi-processed data reveal many obvious results about the degree of change in standard of living of the people. Their income level and housing conditions have considerably improved. They have diversified their occupations adopting more remunerative ones so as to increase their income level.

'Higher level of exposure to mass media is also a surrogate indicator of high standard of living. Earlier they had hardly footwear, and even if they had, they use to wear it only on ceremonial occasions, but now almost every men and women wear at least *chappal,* if not shoes, and better quality shoes on ceremonial occasions. They are comparatively better dressed specially the educated ones who wear pant and colourful shirts. Their food habits and quality of food have also improved. They no longer rely on parched grain with piece of *gur* or salt and a jug of water as breakfast or with two breads with salt and onion as their lunch. The social status has also improved mainly due to their vote bank which now-a-days is not blindly utilized, and partly due to their improved economic condition and overall awareness among them. As a result of all these a little less than two-fifth of the total respondents fall in the high category, while nearly 32 per cent moderate standard of living. A lower percentage of respondents amounting to nearly 29 per cent fall under low level of standard of living.

There are considerable differences in the standard of living of user and non-user respondents because users get more agricultural production and income mainly due to the use of more sprinkler irrigation than their counterpart. Therefore, nearly 47 per cent user respondents belong to higher level of standard of living, while 21 per cent non-user respondents fall under this category. Almost equal number of respondents of user and non-user of sprinkler irrigation have medium standard of living but in case of low standard of living, nearly 22 per cent users fall under it as compared to nearly 29 per cent non-users.

8.3.3 Level of Housing Condition

The house is an individual's identity symbol reflecting ones social prestige and status. A suitable house is one of those basic needs of human life without which one cannot conceive of a life worth living. It influences ones physical health and mental efficiency. Consequently, it affects the future skill and productivity which ultimately determines individuals socio-economic status. Prior to independence or even up to its two decades later, the houses of the rural masses were in bad shape and were mainly made of mud walls and thatched roofs which they could not renovate annually due to lack of financial resources. They use to keep their cow, goats, pigs and other animals, if at all they were there, in one of the corners of the same house. But as the process of development has brought economic prosperity, they have started investing a lot in improving their housing condition. They are gradually changing their mud-thatched houses into mud-tile, brick-tile and ultimately into Reinforced Bricks and Cement (RBC) houses or Reinforced Cement and Concrete (RCC).

Housing is a universally complex commodity and it is difficult to gauge and compare them in totality. Even if they look the same, they are infinitely varied in character and quality, and whatever single measure is adopted, none by itself appears to reflect all variations attributed to house size, location, architectural style, various infrastructural facilities etc. It is even more complex in rural areas. With this view as many as 16 questions related to various components of houses such as size, location, architectural style, storage, number of rooms, courtyard, guestroom, building material, household utensils, toilet, arrangement for livestock etc.

have been taken into consideration. Since, these attributes of the houses are varied and complex and the response of the respondents were varied and different, it was not possible to make comparative assessment of the status of their houses. Hence, judicious scores have been given to responses of the respondents to various attributes of houses. Finally, the scores of the individual respondents with respect to housing condition have been added in order to get their housing score. Based on the housing scores, the respondents have been grouped into three categories (Table 8.10).

Table 8.10: Housing Condition of Respondents

Category	*Score Range*	*No. of Respondents*		*Total*
		User resp.	*Non-user resp.*	
Good	<4.50	153 (46.36%)	39 (28.89%)	192 (41.29%)
Moderate	4.50-6.00	96 (29.09%)	32 (23.70%)	128 (27.53%)
Poor	>6.00	81 (24.55%)	64 (47.41%)	145 (31.18%)
Total		**330 (100%)**	**135 (100%)**	**465 (100%)**

Source: Collected, compiled and computed by the Author.

Table 8.10 shows that the houses of around 41 per cent of the total respondents are good, while nearly 28 per cent come under moderate category. Nearly 31 per cent of respondents have poor housing conditions. This statistical account clearly shows that the overall housing condition of the masses in India has improved. However, there are differences in the housing conditions of the user and non-user respondents. By and large, the users have better housing conditions than non-users.

8.3.4 Level of Exposure to Mass Media

Exposure to mass media is an irrefutable indicator of socio-economic transformation as it reaches the audience more easily and many a times directly, and affects knowledge, attitude, opinion and behaviour of individual. It not only exposes the masses regarding various innovations and technologies in different fields, but it also demonstrates the benefits and achievements of people who have opted them. With this backlog an attempt has been made to assess the level of exposure of the respondents by posing questions regarding various sources of mass media such as radio, transistor, television, newspaper, magazine, fair, exhibition, motion pictures etc. Queries were also made regarding their participation and involvement in them. As the questions and the responses in this connection were of different nature and character, weighted scores are allotted to different components in order to facilitate the comparability of data and its subsequent quantification. The total scores of the individuals have been worked out and they have been grouped into three categories. The resulting figures are given in Table 8.11.

Table 8.11: Level of Exposure to Mass Media of Respondents

Category	*Score Range*	*No. of Respondents*		*Total*
		User resp.	*Non-user resp.*	
Less Exposed	< 2.0	125 (37.88%)	54 (40.00%)	179 (38.49%)
Moderately Exposed	2.0-5.0	99 (30.00%)	43 (31.85%)	142 (30.54%)
More Exposed	> 5.0	106 (32.12%)	38 (28.15%)	144 (30.97%)
Total		**330 (100%)**	**135 (100%)**	**465(100%)**

Source: Collected, compiled and computed by the Author.

A glance at individual score shows that there is lot of variation in it ranging from 1.00 to 7.5 score. But larger number of respondents are on the higher side. Table 8.11 showing the level of exposure of respondents to mass media indicates that nearly 39 per cent respondents are highly exposed to mass media, while about 31 per cent are moderately exposed. The higher level of exposure is due to the fact that radio and transistors have become very common, and television is gaining popularity. The increasing level of literacy, the newspapers and magazines, specially the local ones, are also contributing to the level of exposure. Moreover, tea stalls, *chaupals,* market places and many such other places of gathering are also contributing a lot to the level of exposure because here people not only discuss the political affairs of local, national and international nature, but many innovative technologies and their achievements are also discussed which ultimately enhance their level of exposure. Table 8.11 further shows that there is not much difference in the level of exposure to mass media between user and non-user respondents.

8.3.5 Operational Size of Holding

In our country in general and Haryana in particular, there is a very complex tenurial system, although the State and the Central Governments are trying to simplify it by enacting various legislations from time to time. But unfortunately or fortunately, the Indians are masterminded in finding the loopholes in the law or interpreting the laws in their favour against its spirit. And as such, practically the tenurial system becomes more and more complex, although theoretically it is being simplified and reformed. However, the owners of land, who are not in a position to cultivate their land themselves, or who do not want to cultivate the land, give their land to actual tiller on various terms and conditions. The nature of the term and conditions depends on the supply and demand of land to be leased out. If the demand is more and supply is meagre, the owner of land exploits the tiller by giving his land to him (tiller) according to his term and conditions. On the contrary, when supply of land is more and demand is less, the tillers of land dictate the terms, and the owner has to accept it.

Nevertheless, whatever may be the tenancy system, it is the total area being cultivated by the respondents which matters more than the land registered in his name because it is in this size of holding in which he has to toil and supply all his inputs. Therefore, in the present study an attempt has been made to find out the operational size of holding of each respondents. After ascertaining the operational size of holding of each respondent, they have been arranged in ascending order and finally classified into 5 categories (Table 8.12).

Table 8.12: Operational Size of Holding of Respondents

Size of Holding	*Size range*	*No. of Respondents*		*Total*
		User resp.	*Non-user resp.*	
Marginal	< 1.00 ha	47 (14.24%)	32 (23.70%)	79 (16.99%)
Small	1.0-2.0 ha	54 (16.36%)	37 (27.41%)	91 (19.57%)
Semi-Medium	2.0-4.0 ha	128 (38.79%)	43 (31.85%)	171 (36.77%)
Medium	> 4.00 ha	101 (30.61%)	23 (17.04%)	124 (26.67%)
Total		**330 (100%)**	**135 (100%)**	**465 (100%)**

Source: Collected, compiled and computed by the author.

Table 8.12 represents spatial pattern of the respondents. There are 79 respondents who account nearly 17 per cent respondents, are marginal farmers with less than one hectare of land. There are 91 respondents accounting nearly 20 per cent of the total with small size of holdings, while 191 respondents contributing nearly 37 per cent of the total have semi-medium size of holding. The remaining 8 per cent have medium size of holding of more than 4 hectares.

The spatial pattern of size of holding is different among the user and non-user of sprinkler irrigation. On the whole the users have larger size of holding as compared to non-users. Nearly 31 per cent of the users have marginal and small size of holdings, while it is nearly 50 per cent in case of non-users.

8.3.6 Overall Level of Economic Transformation

The above discussion and analysis of level of economic transformation with respect to selected economic indicators show intra- and inter-indicator variations in all of them. Hence, in order to get the aggregate picture of level of economic transformation the composite index of economic transformation has been worked out using the same formula used in overall social transformation. The result obtained in this equation is 3.29 which has been rounded up to 3.50 and finally three grades, *viz.* more than 3.50 score high, 3.50 to 2.50 medium and less than 2.50 low score have been identified. The respondents have been grouped into three categories (Table 8.13).

Table 8.13: Composite Index of Economic Transformation of Respondents

Category	*Score Range*	*No. of Respondents*		*Total*
		User resp.	*Non-user resp.*	
High	> 3.50	139 (42.12%)	19 (14.07%)	158 (33.98%)
Medium	3.50-2.50	104 (31.52%)	49 (36.30%)	153 (32.90%)
Low	< 2.50	87 (26.36%)	67 (49.63%)	154 (33.12%)
Total		**330 (100%)**	**135 (100%)**	**465 (100%)**

Source: Collected, compiled and computed by the author.

Table 8.13 shows that about 31 per cent of the total respondents have recorded high economic transformation. A closer scrutiny of the master table shows that larger number of respondents (41 respondents) are under this category who comparatively have larger size of holding. About 41 per cent respondents have shown moderate change and they are distributed across all the categories of respondents. Nearly 28 per cent respondents have shown low level of transformation and they are mostly poor respondents.

Table 8.13 further shows the difference in the economic

transformation of users and non-user respondents. Nearly 42 per cent of the user respondents have high level of economic transformation while only 14 per cent non-user respondents come under this category. On the contrary nearly half of the non-users have low level of economic transformation as against 26 per cent user respondents. Almost the same percentage of respondents of both the categories fall under medium category. This also proves our null-hypothesis, viz. "Use of sprinkler irrigation leads to agricultural development which in turn results into economic transformation."

8.3.7 Overall Level of Socio-Economic Transformation

In the proceeding analysis it has been found that there is a lot of variation in the levels of transformation not only in various selected socio-economic indicators, but also in the aggregate of social and economic transformation. Therefore, in order to assess the overall socio-economic transformation a Composite Index of Transformation has been worked out using the following formula:

$$\text{Composite Index of Transformation } (CIT) = \frac{\sum_{x=1}^{n} x_{ij}}{n}$$

where,

x_{ij} = score value of *i*th element of socio-economic change; and

n = number of element of socio-economic change.

The result obtained in this equation is 3.26 which have been rounded to 3.50. Taking one as an interval, three categories, viz. High, Medium and Low have been formed and the respondents have been put in different categories as per their scores (Table 8.14).

Table 8.14: Overall Level of Socio-Economic Transformation

Category	*Score Range*	*No. of Respondents*		*Total*
		User resp.	*Non-user resp.*	
High	> 3.50	117 (34.86%)	35 (25.92%)	150 (32.26%)
Medium	3.50-2.50	129 (39.09%)	76 (56.30%)	205 (44.09%)
Low	< 2.50	86 (26.05%)	24 (17.78%)	110 (23.65%)
Total		**330 (100%)**	**135 (100%)**	**465 (100%)**

Source: Collected, compiled and computed by the author.

Table 8.14 shows that around 32 per cent respondents have overall high level of socio-economic transformation. As has been pointed out earlier that most of these respondents have reasonable size of land holding. Nearly 44 per cent respondents belong to medium level, while around 24 per cent have fallen in the low category. Most of them are poor respondents.

Table 8.14 further shows that in totality the user respondents have higher level of socio-economic transformation than their counterpart. Finally, our proposed hypothesis, i.e. "The use of sprinkler irrigation leads to agricultural development which ultimately results into socio-economic transformations."

The major conclusions which could be derived from the above discussion and analysis are: *firstly,* there are considerable changes in all the socio-economic indicators chosen in this study; *secondly,* the level of change is not uniform in all the cases but there are inter and intra-indicator variations in both social and economic conditions; *thirdly*, there is higher level of transformation in economic conditions than that of social conditions which is in conformity with the general principle, viz. economic prosperity brings social transformation; and *fourthly,* the Composite Index of Transformation also shows high level of transformation. And, all these points prove the proposed hypotheses. There are definitely strong positive changes which are bringing overall improvements in the condition of the masses. Therefore, it is suggested that more and more facilities, incentives, loans and subsidies for various development and welfare programmes should be provided to the people, so that, their pace of progress increases and they could come up at par with the other castes and communities of the nation, but without any kind of confrontation with them.

References

Aggarwal, M.C. *et al.*, 1982: Yield of Bajra and Wheat with Saline Water Applied Through Sprinkler and Surface Irrigation Methods, *Annals Arid Zone*, Vol. 21, No. 1.

Alexender, J.C., 1990: *Culture and Society: Contemporary Debate*, Cambridge University Press, Cambridge.

Asia Development Bank, 2001: *Growth and Change in Asia and the Pacific, (Key Indicators)*, Oxford University Press, Delhi.

Bhatt, J.P., 2003: Prospective Fisheries Development in Uttaranchal State', in Rawat, M.S.S. (ed.), *Central Himalaya: Environment and Development*, Department of Geography, H.N.B. Garhwal University, Srinagar.

Bisht, D.S., 2006: *Poverty, Planning and Developing: A Case Study of Uttarakhand State*, Central Himalayan Institute, Dehradun.

Chakravarti, A., 2001: *Social Power and Everyday Class Relations: Agrarian Transformation in North Bihar.*

Chatto, M.A. *et al.*, 2009: 'Vegetable Gardening' in Masood, S. (ed.), *Peri-Urban Agriculture: Agriculture by Twenty and Beyond*, A.P.H. Pub. Corp., New Delhi.

Chikara, D.V., 2006: *Levels of Agricultural Development in Haryana* Unpublished Ph.D. Thesis, Geography Department, Delhi University, Delhi.

Clark, *et al.*, 2000: Micro-Irrigation Design Consideration for Sandy Soil Vegetable Production System, *Internatioinal Water and Irrigation Review*, Vol. 21, No. 4.

Farro, T.J., 2001: *Social Action System and Synthesis in Sociological Theory,* Preger, Westport.

Goel, E.C. and Rao, B.N., 1980: Evaluation of Sprinkler Irrigation System and Comparison with Surface System, *Haryana Agricultural University Research Journal*, Vol. 10, No. 3.

Hanumantha Rao, C.H. and Linneman, H. (eds.), 1996: Economic Reforms and Poverty Alleviation in India: *Indo-Dutch Studies on Development Alternatives*, Vol. 17.

Hooda, A., 2002: "WTO Agreement on Agriculture and India", in Hooda, A. (ed.), *WTO Agreement and Indian Agriculture,* Social Science Press, Delhi.

Hooda, R.S., 1994: Impact of Sprinkler Irrigation on Crop-combination-A Case Study of Bhiwani District, *Crop Research*, Vol. VII, No. 1.

Joshi, P.C., 1989: *Culture, Communication and Social Change*, Vikas Publishing House, New Delhi.

Kabilan, R., 1993: *From Traditional Society to Post Industrial Society*, Commonwealth Publishers, New Delhi.

Kaushal, M.P., 1976: Use of Sprinkler Irrigation in Sand Dune Areas, *Agriculture and Agro-industries,* Vol. 9, No. 5.

Kingdon, G., 1996: "The Quality and Efficiency of Private and Public Education: A Case Study of Urban India", *Oxford Bulletin of Economics and Statistics*, 58(1), 57-82.

Kuznets, S., 1973: "Modern Economic Growth: Findings and Reflections", *American Economic Review*, 13(3), 247-258.

McGranahan, D., Pizarro, E. and Richard, C., 1985: *Measurement and Analysis of Socio-Economic Development*, N.N. Research Institute, Geneva.

Misger, F.A., Kumar, A., and Masood, S., 2009: 'Water and Fruit Crops' in Masood, S. (ed.), *Peri-Urban Agricultural: Agriculture by Twenty and Beyond*, A.H.P. Pub. Corporation, New Delhi.

Mohammad, N., 1978: Impact of Social Factors on Diffusion of Agricultural Innovations in Central Trans-Ghaghara Plan, *National Geographical Journal of India*, Vol.XXIV, Varanasi.

Mohammad, N., 1988: *Caste and Primary Occupation: A Geographical Analysis*, Concept Publishing Co., New Delhi.

Mohammad, N., and Majeed, A., 1979: "Socio-economic Factors and Diffusion of Agricultural Innovations" in Mohammad, A. (ed.), *Dynamics of Agricultural Development in India*, Concept Publishing Co., New Delhi.

Mohammad, N., 1992: "Spatial Variations in Levels of Development in U.P." in Mohammad, N., (ed.) *Perspectives in Agricultural Geography,* Vol. V, Concept Publishing Co., New Delhi.

Neelofar, L. and Masood, S., 2009: "Floriculture, Landscaping and Recreation" in Masood, S. (ed.), *Peri-Urban Agriculture: Agriculture by Twenty and Beyond*, A.P.H. Publishing Corp., New Delhi.

Puroth, M.L., Saha, D.K. and Bhandari, 1998 : Socio-economic Aspects of use of Sprinkler Irrigation in Sikar District of Rajasthan, *Proceedings of CBPI*; Jalgaon.

Rogaly, B. and Barbara Harriss-White, Bose, S. (eds), 1999: *Sonar Bangla?* Agricultural Growth and Agrarian Change in West Bengal and Bangladesh.

Roy, P.K. and Sharma (eds.), 2010: *Globalisation and Diversification in India*, Regal Pub., New Delhi.

Sathyamurthy, T.V., *Class Transformation and Practical Transformation in Post-Colonial India*, New Delhi.

Shiva, V. and Bedi, G. (eds.), 2002: *Sustainable Agriculture and Food Security: The Impact of Globalisation.*

Siddiqui, M.A.A. and Masood, S., 2009: 'Medicinal and Aromatic Plants' in Masood, S. (ed.), *Peri-Urban Agriculture: Agriculture by Twenty and Beyond*, A.P.H. Pub. Corp., New Delhi.

Singh, N.P. *et al.*, 2004: *Modern Technology on Vegetable Production*, International Book Distributing Co., Lucknow.

Singh, R.P., 2007: "Role of Biofertilizer in Crop Production" in Sharma, A. Singh, R.K. (eds.), *Composite Farming Practice and Economic Development*, Abhijeet Pub., New Delhi.

Sinha, R.R.K., 2009: *Dynamics of Land-Caste Relations in India*, Manak Pub., New Delhi.

Sundaram, K., and Tendulkar, S.D., 1988: *"Measurement of Poverty: Some Conceptual Problems"*, Paper No. 36, Second Seminar on Social Statistics, Feb. 4-6, 1988, Central Statistical Organisation, New Delhi.

Tendulkar, S.D., 1992: 'Economic Growth and Poverty' in Hariss, B., Guhan, S. and Csssen, R. (eds) *Poverty in India: Research and Policy*, Oxford University Press, Bombay. I, Sections III and IV, pp. 32-42.

Thakur, B., 2011: *Depletion and Degradation of Land Resource of India*, R.N. Dubey Foundation, Allahabad.

Tornquist, O., 1999: *Politics and Development: A Critical Introduction,* Uppsala University.

Verma, K., 2007: Role of Urban Agriculture in Life Support System, A Case Study of Alipur Block, Delhi (Unpublished M.Phil. Dissertation, Department of Geography, Delhi University, Delhi).

World Bank, 1997: *World Development Report-1997*, Washington D.C., UK.

9

Summary, Conclusion and Suggestions

The State of Haryana came into existence on November 1, 1966. Covering an area of 44,212 km², it extends from 27° 39′ to 30° 55′ N. Lat. and 74° 27′ to 77° 38′ E. Longitude. A greater part of Haryana falls in the Indo-Ganga divide and the remaining smaller portion is covered with Siwalik ranges and Aravalli system. Nearly half of its south-western part experiences arid to semi-arid conditions with scanty and highly unreliable rainfall which leads to the occurrence of droughts and famine. The south-western Haryana has low water potential and agriculture is largely dependent on irrigation which has considerably affected the agricultural landuse and its productivity. Therefore, south-western Haryana has been studied in the study.

The thirsty region of South-western Haryana lies between 27° 47′ and 29° 16′ N. Latitudes, and 75° 8′ and 76° 23′ E. Longitudes and is highly susceptible to drought and famine. It is a part of Indo-Ganga divide which lies between Peninsular India in the south and recently built Himalayan mountain in the north. The region belongs to the recent period of geological history and is formed by the tectonic movements which created a sag between the northern edge of Gondwana land and recently built Himalayan chain. The shallow depression is filled by the sediments brought by the Himalayan rivers in the north and Peninsular India in the south. It ranges in age from the Pliestocene to the recent times. These are of varying thickness and consists of silt and clay with occasional gravel beds. Geomorphologically, it is divided into two major divisions, i.e. Indo-Ganga plain and Aravali system. The former covers substantial portion of the region and is divided into the Bhangar land, and sandy undulating plain, while the latter covers small portion lying in southern part in north-south direction. The Bhangar lands lying in the northern part of the region are older deposits of Pleistocene age. They are generally found on higher ground and are free from floods. The Bhangar land usually ends in broken chain of sand dunes and cliffs.

In the southern part of the region, the geological events of Aravali synclinorium are exposed. The lithological units are mainly of Pre-Cambrian quartzite, schist, mica, phillies etc. They run from south-east to north-west with a series of outcrops of different sizes.

Physiographically, south-western Haryana is divided into 4 physical divisions, i.e. the alluvial plain, the sand dunes, the undulating sandy plain and the rocky hills. The alluvial plain is found in the north, while the sand dunes lie in the western part. The eastern and southern portion is covered by the undulating sand plain, whereas the rocky hills are scattered in north-south direction. The region has a monotonous gentle plain with Aravali and its outcrops scattered over it. There is almost absence of major rivers.

The climate of the region is arid to semi-arid and is characterised by a lot of variations in temperature, rainfall and relative humidity. The variable nature of climatic parameters have a strong bearing on pattern of landuse specially of agricultural landuse and cropping system. These conditions prevail mainly due to its continental location, distance from mountain and sea. However, climate of South-western Haryana may be briefly summarized as a sub-tropical continental monsoon type having characteristics of seasonal rhythm, hot summer, cold winter and great annual range of temperature.

The distribution of rainfall is highly irregular both in space and time and water deficiency is a problem. There is marked difference in the distributional pattern of annual rainfall due to prevalence of easterly moving monsoon and westward advancing westerly disturbances. The average annual rainfall is less than 300 mm in north-western part, while in eastern part it is more than 300 mm. There are considerable variations in the seasonal and annual rainfall. The co-efficient of variability of annual rainfall is 30 to 50 and its distributional pattern is just opposite to distributional pattern of rainfall. The average annual potential evaporation is around 143 cm which decreases towards east, while the average annual water deficit is around 110 cm which also decreases from south-west towards north-east.

The region has a variety of soils having a complex mosaic distributional pattern which is reflected at every level. It is mainly due to a variety of parent materials, variable amount of rainfall, intensity of temperature and local as well as regional pattern of soil formation. The variations in anthropogenic determinants like intensity of irrigation, landuse, hydrological character have strong bearing on pattern of soil distribution in the region.

However, the soils of south-western Haryana may be classified into 5 types based on their nomenclature. Nearly 50 per cent area of the region is covered with coarse sandy soil which consists of large size of sand particles and is dotted with sand dunes. They are subjected to excessive wind erosion and are very loose soil with maximum permeability causing excessive drainage and low water holding capacity. The loamy sandy soil has a fairly good structure and texture and is interrupted with variety of dunes with various sizes and shapes. These are porous soils with moisture deficiency. The sandy loam soil consists of larger size of sand particles and gravel of different sizes. The fine loamy soil is soft soil with almost ideal proportion of sand, silt and clay. The rocky surface and shallow piedmont are shallow, coarse, and infertile.

South-western Haryana supports a population of 2,62,222 with an average density of 336 persons per km^2. It is predominantly inhabited by rural population. The literary level is considerably high with 67 per cent as literate population. The region has basically an agrarain economy engaging nearly 82 per cent of total population directly and indirectly in it. About 82 per cent of the total area is culturable. Total cropped area amounts to 9,72,500 ha giving 188 per cent intensity of cropping. Forest cover is almost negligible, but net sown area covers nearly 82 per cent of total area, which shows that there is not much scope for further horizontal expansion of agriculture. Fallow land also occupy a small percentage, while land not available for cultivation covers more than 12 per cent of total area.

The agricultural year is divided into two seasons, i.e. *kharif* and *rabi*. The major crops of the *kharif* season are *jowar*, *bajra*, pulses, cotton, sugarcane etc., while wheat, barley, gram, pulses, oilseeds are the main crops of rabi season. *Bajra* occupying nearly 58 per cent of net sown area or more than 30 per cent of gross cropped area is the dominant crop of the region. It is grown in all the community development blocks. Wheat is next important crops which covers 29 per cent of net sown area and about 15 per cent of gross cropped area. It is grown all over the area, but higher concentration is in Hisar and Mahendragarh tahsils. Pulses including gram occupy nearly 42 per cent of the net sown area and are grown all over the region, but higher concentration is in Tosham and Dadri tahsils.

Many varieties of oilseeds are grown in the region. They altogether occupy about 14 per cent of net sown area. Besides these, there are several other crops grown in the region. The other crops occupy more than one-fourth of the net sown area.

As the agriculture is not much diversified in the region, few crops-combination regions have emerged in the study. Three regions of 2 crops-combinations and 3 crops-combinations each have emerged, while 2 regions of 4 crops-combination have been identified. However, there is not a single region with the same crops in same sequence. The crops have kept on changing, but only 7 crops have come in combination. *Bajra* has come in all combinations while wheat has appeared in majority of cases.

Although during the last three decades, the agriculture has developed, but it has not been able to keep pace with the state average. The analysis of 4 sets of indicators indicates that much development has taken place in agricultural development. It indicates that none of the tahsils fall under high level of agricultural development. However, 3 tahsils, viz. Hisar, Bhiwani and Tosham have achieved medium level of development, while Dadri and Mahendragarh tahsils have gone to low level of development. The remaining three tahsils, viz. Siwani, Loharu and Narnaul have very low level of development. The comparison of different indicators of agricultural development confirms the overall level of development in the region. A closer scrutiny of use of sprinkler irrigation and level of agricultural development shows that by and large, the users of sprinkler irrigation have higher level of agricultural development as compared to their counterpart. And, this proves our first null-hypothesis, viz. "The sprinkler irrigation has positive correlation with agricultural development".

Age composition of the population is very important parameter in the process of overall development in general and rural development in particular. It ascertains the workforce and level of dependency. The favourable age composition is also instrumental in the adoption of new ideas and practice which bring economic benefits and prosperity. Hence, if the leadership to household is in the hands of energetic younger generation, the probability of prosperity and overall change is more.

The study reveals that the younger generation of less than 30 years of age are coming forward in the decision-making process of the household matters specially related with the economic issues. Nearly 32 per cent population belongs to younger generation and the other 27 per cent to middle age-group which was not the case even two decades before. They will accelerate the process of transformation due to their higher level of education, more mobility, and interaction as compared to elder generation who resist to change from their age old practices, and adoption of innovations. Therefore, the younger generation specially the educated ones should be encouraged to take the lead in the decision-making process of the household matters. The analysis further reveals that the user of sprinkler irrigation are much large in number as compared to their counterpart in the younger age group.

Education is, of course, unfailing engine of change and is the chief instrument through which it socialises its members for bringing desirable changes. It is the aggregate of all social process through which people develop abilities, attitudes and other form of behaviour of positive values. Hence, special focus has been given in this study on educational status of the people, preference of the school and its location for both male and female child. The study reveals that the educational status of the people respondents is not very encouraging but the users of sprinkler irrigation are more literate than non-users. By virtue of being more literate, they have adopted sprinkler system of irrigation.

The situation is a little more encouraging when the educational status of the households is ascertained by working out the Index of Educational Status wherein the educational score of all the household members have been added together and divided by all household members excluding the children of less than 5 years of age. The result shows that 28 per cent households are less educated, while 26 per cent are highly educated. The remaining 46 per cent are moderately educated.

The size of family is another social parameter, which has undergone a change across the board. Earlier the large size of family was considered as a measure of social security and were instrumental income generation, but now they have realized the merits and advantages of small and medium size of family which is reflected by the fact that about 60 per cent people are covered in these two categories. About 38 per cent people have large and very large size of family. Hence, it is suggested that rural population still need more education and counselling regarding in the size of families of user and non-user of sprinkler irrigation. The former are in a better position than latter.

The size of family is very much associated with the adoption of family planning programme. The study reveals that the educated younger people are in favour of changing the structure and system of their family. In the ultimate statistical analysis, about 41 per cent people have gone to the higher category, while nearly 30 per cent belong to medium category. Generally, users of sprinkler irrigation have political awareness and involvement which is another important indicator of social transformation. The study reveals that the democratic process of elections at various levels has created awakening, interest and involvement among the rural masses. The consolidated processed result indicates that more than one-third population is more aware and involved in political affairs, while about 32 per cent are moderately involvea which is a positive sign of social change. Almost similar results have been obtained in case of scheduled castes also, but the non-scheduled caste population is a little more involved because of their socio-economic stand. The political involvement increases the linkage of the masses with politicians, officer and decision-makers which ultimately helps in their socio-economic well-being. Hence, the masses should be more motivated and encouraged in political affairs, but not in politic king.

Moreover, it has also been found that, by and large, the users of sprinkler irrigation have an edge over non-users which is another indicator of social change. Therefore, their composite index of social transformation is higher than non-users.

However, sprinkler irrigation brings social transformation. The composite index of overall social transformation shows that nearly 40 per cent of the population belongs to high degree of change with more than 3.5 score, while about one-third come under medium category. As such only 27 per cent fall under low category. Hence, the overall hypothesis of social transformation is proved, viz. "Sprinkler irrigation brings social transformation".

Various programmes and policies of rural development in general and agricultural development in particular have resulted into economic prosperity of the masses.

Income is the main basis for transformation in other economic parameters also such as standard of living, housing condition, exposure to mass media and so on. The index of standard of living indicates that nearly 38 per cent population belongs to high category, while 30 per cent fall under medium category. As such 32 per cent have low standard of living. It is, therefore, suggested that whatever schemes for the upliftment of the rural masses are introduced, they should be properly implemented.

The house is the identity symbol of our social prestige and status. A suitable house influences the physical health, mental efficiency, future skill and productivity which determines the socio-economic status. After Independence there has been tremendous improvement in the housing condition across the board in our country. Scheduled castes are spending a lot on the housing. The size, structure, building material, amenities and so on have undergone tremendous change. The index of housing condition based on scoring schemes shows that about 41 per cent population with more than 4.5 scores live with good housing condition. Nearly 28 per cent people have moderate housing condition, while 31 per cent have poor houses. This is a very encouraging picture indicating economic change. The housing condition of user respondents is much better than non-users. Moreover, large number of users have better housing condition than non-users of sprinkler irrigation. Hence, it is suggested that not only more and more housing schemes should be introduced but they should be properly implemented.

Exposure to mass media affects the knowledge, attitudes, opinion and behaviour, which in turn affects levels of innovations and technology leading to positive economic change. Based on index of change, the study reveals that nearly 38 per cent of population are more exposed, while nearly 31 per cent are moderately exposed. Only 31 per cent are less exposed which shows that radio, television, newspaper, periodicals etc. have played important role in exposing the masses with the latest development in various fields not only at national level, but also at international levels. However, there is not much difference in the level of exposure of users and non-users of sprinkler irrigation.

The standard of living is a very effective indicator to gauge the socio-economic condition of the people. Although, mere observation and making qualitative comments of standard of living is easier, but its qualitative and quantitative measurement is comparatively difficult because it is, by and large, manifestation of income, housing condition, dress and dressing materials, household utensils, exposure to mass media, social status and a host of such conditions. Per capita income is, perhaps, the main factor, but it is not everything for the assessment of standard of living.

The analysis and observation of the collected facts and processed data reveal many obvious results about the degree of change in standard of living of the people. Their income level and housing conditions have considerably improved. They have diversified their occupations adopting more remunerative ones so as to increase their income level. Higher level of exposure to mass media is also a surrogate indicator of high standard of living. Earlier they had hardly footwear, and even if they had, they use to wear it only on ceremonial occasions, but now almost every men and women wear at least *chappal*, if not shoes, and better quality shoes on ceremonial occasions. They are comparatively better dressed specially the educated ones who wear pant and colourful shirts. Their food habits and quality of food have also improved. They no longer rely on parched grain with piece of *gur* or salt and a jug of water as breakfast or with two breads with salt and onion as their lunch. The social status has also improved mainly due to their vote bank which now-a-day is not blindly utilized, and partly due to their improved economic condition and overall awareness among them. As a result of all these, two-fifth people fall in the high category, while nearly 32 per cent have moderate standard of living. A lower percentage of respondents amounting to nearly 29 per cent fall under low level of standard of living.

There are considerable differences in the standard of living of user and non-user population of sprinkler irrigation because users get more agricultural production and income mainly due to the use of more sprinkler irrigation than their counterpart.

In our country in general and Haryana in particular, there is a very complex tenurial system, although the State and the Central Governments are trying to simplify it by enacting various legislations from time to time. But unfortunately or fortunately, the Indians, are masterminded in finding the loopholes in the law or interpreting the laws in their favour against its sprit. And as such, practically the tennurial system becomes more and more complex, although theoretically is being simplified and reformed. The analysis of operational size of holding records that people accounting to nearly 17 per cent population are marginal farmers with less than one hectare of land. There are 91 persons accounting nearly 20 per cent of the total with small size of holdings, while 191 people contributing nearly 37 per cent of the total have semi-medium size of holding. The remaining 8 per cent have medium size of holding of more than 4 hectares.

The spatial pattern of size of holding is different among the user and non-user of sprinkler irrigation. On the whole the users have larger size of holding as compared to non-users. Nearly 31 per cent of the users have marginal and small size of holdings, while it is nearly 50 per cent in case of non-users.

The analysis level of economic transformation with respect to selected economic indicators show inter and intra indicator variations in all of them. The results of composite index of economic transformation shows that about 31 per cent of the total population have recorded high economic transformation. A closer scrutiny of the master table shows that larger number of people coming under this category are 41 who comparatively have larger size of holding. About 41 per cent respondents have shown moderate change and they are distributed across all the categories of people. Nearly 28 per cent people have shown low level of transformation and they are mostly poor masses.

Table 8.13 further shows the economic transformation of users and non-user respondents. Nearly 42 per cent of the user population have high level of economic transformation while only 14 per cent non-user respondents come under this category. On the contrary nearly half of the non-users have low level of economic transformation as against 26 per cent user respondents. Almost same percentage of respondents of both the categories fall under medium category. This proves our hypothesis, viz. "the use of sprinkler irrigation results into economic transformation".

The results of overall level of socio-economic transformation shows that around 32 per cent people have overall high level of socio-economic transformation. As has been pointed out earlier that most of these people have reasonable size of land holding. Nearly 44 per cent people belong to medium level, while around 24 per cent have fallen in the low category. Most of them are poor people. Table 8.13 further shows that in totality the user population have higher level of socio-economic transformation than their counterpart. The critical analysis of socio-economic transformation proves the hypothesis, i.e. "use of sprinkler irrigation leads to socio-economic transformation".

The major conclusion which could be derived from the above discussion and analysis are: *firstly*, there are considerable changes in all the socio-economic indicators chosen in this study; *secondly*, the level of change is not uniform in all the cases, but there are inter and intra-indicator variations in both social and economic conditions; *thirdly*, there is higher level of transformation in economic conditions than that of social conditions which is in conformity with the general principle, viz. economic posterity brings social transformation; and *fourthly*, the Composite Index of Transformation also shows high level of transformation. And, all these points prove the proposed hypotheses. There are definitely strong positive changes which are bringing overall improvements in the condition of the masses. Therefore, it is suggested that more and more facilities, incentives, loans and subsidies for various development and welfare programmes should be provided to the people so that their pace of process increases and they could come up at par with the developed group of the population of the nation but without any kind of confrontation with them.

Bibliography

Abraham, T. and Lamathi, E. 'Eco-organic Farming' in Pruhit, S.S. and Gehlot, D.C. (eds.) *Trends in Organic Farming in India*, Jodhpur.

Aggarwal, M.C. *et al.*, 1982: Yield of Bajra and Wheat with Saline Water Applied Through Sprinkler and Surface Irrigation Methods, *Annals of Arid Zones*, Vol. 21, No. 1.

Aggarwal, M.C., 1985: Sprinkler Irrigation: A Case Study in Haryana, *Indian Agricultural Research Institute*. New Delhi.

Agrews, C., 2002: Drought, Desertification and Desiccation: Need for Further Analysis, *Geography*, Vol. 87, No. 3.

Agricultural Statistical at a Glance, 2006-2012. Directorate of Economics and Statistics, Ministry of Agriculture, New Delhi.

Ahmad, M.S., 2004: *Five Decades of Planning and Developing: A Case Study of Uttarakhand.*

Alagh, Y.K. (ed.) 2003: *Globalisation and Agricultural Crisis in India*, Deep & Deep Publication, New Delhi.

Alexender, J.C., 1990: *Culture and Society: Contemporary Debate*, Cambridge University Press, Cambridge.

Anonymous, 1998: Sprinkler Irrigation in India, *INCID*, New Delhi.

Asia Development Bank, 2001: *Growth and Change in Asia and the Pacific, (Key Indicators)*, Oxford University Press.

Athewala, A.G., 1966: Some New Methods of Crop Combination, *Geographical Review of India,* Vol. XXXVIII, No. 1, Kolkata.

Auden, J.B and Roy, P.C., 1942: Report on the Sodium Salt-Reh in U.P., *Records* of *Geological Survey of India*, Calcutta.

Ayyar, N.P., 1969: Crop Regions of Madhya Pradesh, *Geographical Review India*, Vol. XXXI, No. 1, Kolkata.

Bader, M.D.M., 2010: *Disparities in Neighbourhood for Food Environment: Implications, Measurement and Strategies.*

Bhatia, S.S., 1965: Patterns of Crop Concentration and Diversification in India, *Economic Geography*, Vol. 43.

Bhatt, J.P., 2003: "Prospective Fisheries Development in Uttaranchal State", in Rawat, M.S.S. (ed.), *Central Himalaya: Environment and Development*, Department of Geography, H.N.B. Garhwal University, Srinagar.

Bhattarai, N. and Narayanmoorthy, 2003: Impact of Irrigation on Rural Poverty in India: An Aggregate Panel Data Analysis, *Water Policy*, Vol. 5, Nos. 5-6.

Bisht, D.S., 2006: *Poverty, Planning and Developing: A Case Study of Uttarakhand State*, Central Himalayan Institute, Dehradun.

Blanford, H.F., 1889: *Climate and Weather of India, Ceylon and Burma*, London.

Borlaug, N.E., 2004: Feeding a World of 10 Billion People: Our Twentyfirst Century Challenge, in Scaners, C.G. and Miranoski, I.A. (eds.) *Perspectives in World Food and Agriculture,* Blackwell, Two.

Brerno Dorin and Frederi, L. (eds.), 2009: *Agriculture and Food in India*, Manohar Publishers, New Delhi.

Butola, B.S., 2003: Drought on Doom: Its story of Rajasthan, *Journal of Geography*, Vol. 3, No. 5.

Census of India, 2011: *Haryana, Administrative Atlas*, Directorate of Census.

Census of India; 2001: *Haryana-Administrative Atlas*, Directorate of Census.

Chakravarty, A., 2001: *Social Power and Everyday Class Relations: Agrarian Transformation in North Bihar.*

Chatty, M.A. *et al.*, 2009: 'Vegetable Gardening' in Masood, S. (ed.), *Peri-Urban Agriculture: Agriculture by Twenty and Beyond*, A.P.H. Pub. Corp., New Delhi.

Chikara, D.V., 2006: *Levels of Agricultural Development in Haryana,* Unpublished Ph.D. Thesis, Geography Department, Delhi University, Delhi.

Clark, *et al.*, 2000: Micro-Irrigation Design Consideration for Sand Soil Vegetation Production System, *International Water and Irrigation Review*, Vol. 19, No. 4.

CWC, 1905: Water and Related Statistics, *Central Water Commission*. Government of India, New Delhi.

Dagal, K.C., 1997: Defining Droughts in Marginal Areas: The Role of Perception, *Professional Geographer*, Vol. 49, No. 2.

Dash, N.K., 1998: *Drip Irrigation and Micro-Irrigation Potential and Prosperity for Sugarcane Cultivation with Reference to Orissa*, VSI.

Department of Economics and Statistical Analysis; 2006:

Desai, B.M., 2003: Drought Impact and Vision of Proofing, *EPW*, Vol. 38, No. 21

Desai, D.N. *et al.*, 1993: *Feasibility of Mini-sprinkler in Onion*, Central Board of Irrigation and Power (CBIP).

Dhawan, B.D., 1988: Role of Irrigation in Raising Intensity of Cropping, *Journal of Indian School of Political Economics*, Vol. 3, No. 4.

Dhawan, B.D., 2002: *Technological Change in Indian Irrigated Agriculture: A Study of Water Saving Methods*, Commonwealth Publishers, New Delhi.

Director of Economics and Statistics: *Indian Agricultural Statistics*, Vol. I & II, New Delhi.

Directorate of Economics and Statistics, Ministry of Agriculture, *Agricultural Statistics at a Glance,* 2005 to 2012.

Doi, K., 1959: The Industrial Structure of Japanese Prefecture, *Proceeding of International Geographical Union,* 1957 Regional Conference in Japan.

Duggal, S.L., 1975: *Soil Geography of Haryana,* Hisar Agri. University, Hisar.

Farro, T.J., 2001: *Social Action System and Synthesis in Sociological Theory,* Preger, Westport.

Gajbheye, K.S. and Mandal, C., 2000: Agro-ecological Zones: Their Soil Resources and Cropping System. *Status of Farm Management in India*, New Delhi.

Goel, E.C. and Rao, B.N., 1980: Evaluation of Sprinkler Irrigation System and Comparison with Surface System, *Haryana Agricultural University Research Journal*, Vol. 10, No. 3.

Gohring, T.R. and Wallender, W.W., 1987: Economics of Sprinkler Irrigation System, *Agricultural Economics*, Journal of American Society, No. 30, No. 4.

Goswani, S.N. *et al.*, 2002: Impact of Irrigation on Landuse Efficiency and Area Allocation in Different Class Size of Landholdings in Maharashtra, *Geographical Review of India*, Vol. 64, No. 4. Kolkata.

Government of India, 1999: *Report of Working Group on Water Availability for Use*, National Commission for Integrated Water Resource Development Plan, New Delhi.

Gregor, H.F., 1970: *Geography of Agriculture: Themes in Research*, Prentice Hall.

Hanumantha Rao, C.H. and Linneman, H. (eds.), 1996: Economic Reforms and Poverty Alleviation in India: *Indo-Dutch Studies on Development Alternatives*, Vol. 17.

Hepase, *et al.*, 1993: Micro-Irrigation in Sugarcane Cultivation., *CBIP*.

Hetner, A., 1947: *General Geography of the Humans*, Vol.1, Kohlhammer.

Hooda, A., 2002: "WTO Agreement on Agriculture and India", in Hooda, A. (ed.), *WTO Agreement and Indian Agriculture,* Social Science Press, Delhi.

Hooda, R.S., 1996: Diffusion of Sprinkler Irrigation in Arid India, *Crop Research*, Vol. XI, No. 2.

Hooda, R.S., 1994: Impact of Sprinkler Irrigation on Crop Combination: A Case Study of Bhiwani District, Haryana, *Crop Research*, Vol. VIII, No. 1.

Hooda, R.S., 1994: Role of Capital on Adaptation of Sprinkler Irrigation at Different Size of Holdings in Bhiwani District, *Crop Research,* Vol. VIII, No. 2.

Hooda, R.S., 1990: Sprinkler Irrigation and Agricultural Transformation in Bhiwani District, Haryana, *Geographical Review of India*, Vol. 52, No. 3, Kolkata.

IMCID, 1998: *Sprinkler Irrigation in India,* New Delhi.

Iyer, R.R., 2003: *Water: Perspectives, Issues, Concern*, Sage Publications, New Delhi.

Jain, S.K., Pushpendra, A.K. and Vijay, P.S., 2007: *Hydrology and Water Resources of India*.

Joshi, P.C., 1989: *Culture, Communication and Social Change*, Vikas Publishing House, New Delhi.

Kabilan, R., 1993: *From Traditional Society to Post-Industrial Society*, Commonwealth Publishers, New Delhi.

Karanth, G.K., 1991: Farmers Survival Strategies in a Drought-prone District: A Village in Chiradurga district, *ADRT, Institute of Social Change*, Bangalore.

Kaushal, M.P., 1976: Use of Sprinkler Irrigation in Sand Dune Areas, *Agriculture and Agro-Industries,* Vol. 9, No. 5.

Kazi, M.F. and Wodeyar, A.K., 2001: Agriculture in Flood Prone Areas: A Case Study of North Central Region of Bangladesh, *Indian Journal of Regional Science*, Vol. 33, No. 1.

Kendall, M.G., 1939: The Geographical Distribution of Crop Productivity in England, *Journal of Royal Statistical Society,* Vol. 162.

Kendrew, W.A., 1961: *The Climates of the Continents,* Oxfords University Press, London.

Kingdom, G., 1996: "The Quality and Efficiency of Private and Public Education: A Case Study of Urban India", *Oxford Bulletin of Economics and Statistics*, 58(1), 57-82.

Klages, K.H.W., 1958: *Ecological Crop Geography*, McMillan Co., New York.

Kostrowicki, J., 1972: "Some Method and Techniques to Determine Crop and other Landuse Combinations as Used in Polish Landuse Studies", in M., Shafi, M., Anas and M.F., Siddiqui, (eds.) *Symposium on Landuse in Developing Countries, 21st International Geographical Congress*, Department of Geography, A.M.U., Aligarh.

Krishnan, M.S. 1960: *Geology of India and Burma,* Madras.

Kulkarni, S.A., 2005, *Looking Beyond Eight Sprinklers*, G.B. Pant University, Pantnagar.

Kumar, A., 2002: Water Supply Scheme in Vidarbha, *F.P.W.*, Vol. 42.

Kumar, V., and Pal, M., 1985: Development of Irrigation in Haryana: Limitation and Prospects: A Review, *Transaction of Institute of Indian Geographer*, Vol. 7, No. 2.

Kuznets, S., 1973: "Modern Economic Growth: Findings and Reflections", *American Economic Review*, 13(3), 247-258.

Mahmood, A., 1977: *Statistical Methods in Geographical Studies,* Rajesh Pub, New Delhi.

Majeed, A., 1992: "Crop Combinational Analysis: A Review of Methodology", in Mohammad, N. (ed.) *New Dimensions in Agriculture Geography,* Volume 5, Concept Publishing Co., New Delhi.

McGranahan, D., Pizarro, E. and Richard, C., 1985: *Measurement and Analysis of Socio-Economic Development*, N.N. Research Institute, Geneva.

McHarg, 1966: 'Ecological Determinism' in Darley, F.F. and Melton, J. (eds.) *Future Environment of North American*, New York.

Mehta, H.M. *et al.*, 1993: Comparative Study of Sprinkler *Vs* Flood Irrigation in Summer Groundnut, in Central Board of Irrigation and Power (CBIP).

Memon, K.A., 1995: Micro-Irrigation: Profits in the Pipeline, *India Today*, May, 1995.

Ministry of Environment and Forest, 2010: *National Wetland Atlas of Haryana*. Ahmedabad.

Misger, F.A., Kumar, A., and Masood, S., 2009: 'Water and Fruit Crops' in Masood, S. (ed.) *Peri-Urban Agricultural: Agriculture by Twenty and Beyond*, A.H.P. Pub. Corporation, New Delhi.

Mohammad, A., Munir, A., and Rehman, A., 2007: Fifty Years of Indian Agriculture, Concept Publishing Co., New Delhi.

Mohammad, N. and Arora, M., 2011: Role of Urban Agriculture in Food Security in North-Western Delhi, in Rana, N.K. and K.N. Singh (eds.) *Holistic Development: A Geographical Perspective*, R.K. Book, New Delhi.

Mohammad, N. and Majeed, A., 1979. "Socio-economic Factors and Diffusion of Agricultural Innovations" in Mohammad, A. (ed.), *Dynamics of Agricultural Development in India*, Concept Publishing Co., New Delhi.

Mohammad, N. and Rai, S.C. (eds.), 2013: *Agricultural Diversification and Food Security in Uttarakhand*, Concept Publishing Co., New Delhi.

Mohammad, N. and Raza, M., 1992: "Agricultural Metrological Data Base and Agricultural Productivity" in Mohammad, N. (ed.) *New Dimension in Agricultural Geography*, Concept Publishing Co., New Delhi.

Mohammad, N. Singh, R.B., and Datta, A. (eds.) 2007: *Spatial Information Technology for Natural Resource Management*, Concept Publishing Co., New Delhi.

Mohammad, N., 1975: Crop Association Regions in Ghaghara Rapti Doab, *The North Eastern Geographer,* Vol. VII.

Mohammad, N., 1978. Impact of Social Factors on Diffusion of Agricultural Innovations in Central Trans-Ghaghara Plan, *National Geographical Journal of India*, Vol. XXIV, Varanasi.

Mohammad, N., 1978: *Agricultural Landuse in India*, Inter-India Pub. New Delhi.

Mohammad, N., 1982: "Levels of Agricultural Development in U.P." in Mohammad, N (ed.) *Perspective in Agricultural Geography*, Vol. 4 Concept Publishing Co., New Delhi.

Mohammad, N., 1988: *Caste and Primary Occupation: A Geographical Analysis*, Concept Publishing Co., New Delhi.

Mohammad, N. (ed.), 1996: *New Dimensions in Agricultural Geography*, Vol. I-VIII, Concept Publishing Co., New Delhi.

Nag, P., 2010: *Atlas of Agricultural Resources of India*, NATMO, Kolkata.

Nagrajan, R., 2003: *Drought: Assessment, Monitoring, Management and Resources Conservation.* Capital Publishing Co., Delhi.

Narayanmoorthy, A. and Deshpande, R.S., 2003: Irrigation Development and Agricultural Wages: An Analysis Across States, *E.P.W.*, Vol. 38, No. 35.

Narayanmoorthy, A. 1995: Beneficial Impact of Drip Irrigation: A Study Based on Western India, *Water Resource Journal*.

Narayanmoorthy, A., 2001: Impact of Drip Irrigation on Sugarcane Cultivation in Maharashtra, *Agro-economic Research Centre*, Pune.

Narayanmoorthy, A., 2004: Drip Irrigation in India: Can it Solve Water Crisis, *Development Policy Review*, Vol. 22, No. 4.

Narayanmoorthy, A., 2004: Drip Irrigation in India: Can it Solve Water Scarcity, *Water Policy*, Vol. VI, No. 2.

Neelofar, L. and Masood, S., 2009: "Floriculture, Landscaping and Recreation" in Masood, S. (ed.), *Peri-Urban Agriculture: Agriculture by Twenty and Beyond*, A.P.H. Publishing Corp., New Delhi.

Pascoe, E.H., 1973: *A Manual of Geology of India and Burma-III*, Government of India, Delhi.

Patra, P., 2006: *Land Degradation in Tropical Highlands*, Rajesh Pub., New Delhi.

Pretty, J., 2006: *Ecological Approach to Agricultural Development*, Santiago.

Puroth, M.L., Saha, D.K. and Bhandari, 1998 : Socio-economic Aspects of use of Sprinkler Irrigation in Sikar District of Rajasthan, *Proceedings of CBPI*; Jalgaon.

Rafiullah, S.M., 1965: "A New Approach to Functional Classification of Towns", *The Geographer*, Vol. XII, pp. 40-53.

Rajgopal, A., 1998: Sprinkler System of Irrigation for Water Efficiency in Scarcity Areas, in Verma, V.V.I. and Rao, A.R.G. (eds.): *Micro-Irrigation and Sprinkler Irrigation Systems*, New Delhi.

Ramachandran, R., 1992: "Correlates of Innovations Among Farmers in the Ganga Plain; in Mohammad, N. (ed) *Human Perception on Technological Change in Agriculture,* Concept Publishing Co., New Delhi.

Rao, C.H.H., 2002: "Sustainable use of Water for Irrigation in Indian Agriculture" in Raj Kapila and Unna Kapila (eds.) *Economic Development in India*, Vol. 59, New Delhi.

Raychaudhuri, S.P. *et al.*, 1963: *Soil of India,* New Delhi.

Rogaly, B. and Barbara Harriss-White, Bose, S. (eds.), 1999: *Sonar Bangla?* Agricultural Growth and Agrarian Change in West Bengal and Bangladesh.

Rosergrant, *et al.*, 2002: *World Water and Food to 2020*, International Water Management Institute, Colombo.

Rosergrant, W.M., 1997: Water Resources in Twenty-First Century: Challenges and Implication for Action, *International Food Policy Research Institute*, Washington, DC.

Roy, P.K. and Sharma (eds.), 2010: *Globalisation and Diversification in India*, Regal Pub., New Delhi.

Saleth, R.M., 1996: *Water Institutions in India: Economics, Law and Policy*, Commonwealth Publishers, New Delhi.

Sanklayan, P.L. and Singh, I.P., 1985: Impact of Surface and Lift Irrigation System on Cropping Pattern, *Economic Affairs,* India.

Sathyamurthy, T.V., *Class Transformation and Practical Transformation in Post Colonial India*, New Delhi.

Seckler, D., Banker, R. and Amar Singh, U.A., 1999: *International Journal of Water Resources Development,* Vol. 15, No. 2.

Seckler, *et al.*, 1998: World Water Demand and Supply 1990, to 2025: Scenario and Issue, *Research Report*: 19, International Water Management Institute, Colombo.

Semhadri, S. 1997: Andhra Pradesh: River Water Politics in Drought-prone Telangana, *EPW*, Vol. 32.

Sen, A.K., 2000: *Development as Freedom*, Oxford University Press, New York.

Sengupta, A., 2002: Water Resources and Agricultural Productivity in Malda District, *Geographical Review of India*, Vol. 64, No. 3, Kolkata.

Shafi, M., 1990: Agriculture and Rural Transformation in India, *National Geographer,* Vol. 38, No. 1.

Shafi, M., 2006: *Agricultural Geography*, Pearson Pub., New Delhi.

Shah, S.L., 1985: "Some Methodological and Pragmatic Issues in Land Capability Classification in Himalaya", in J. Singh (ed.), *Environmental Regeneration in the Himalaya: Concept and Strategies,* Gyanodaya Prakashan, Nainital.

Shiva, V. and Bedi, G. (eds.), 2002: *Sustainable Agriculture and Food Security: The Impact of Globalisation.*

Siddiqui, M.A.A. and Masood, S., 2009: Medicinal and Aromatic Plants' in Masood, S. (ed.), *Peri-Urban Agriculture: Agriculture by Twenty and Beyond,* A.P.H. Publishing Corp., New Delhi.

Singal, A.M., 1995: "Approaches to Water Resource Development in Arid Areas of Haryana" in Chauhan, T.S. (ed.) *Irrigation in Arid Lands*, Print Well Publishers, Jaipur.

Singh, H.P. *et al.*, 2001: *Micro Irrigation CBIP*, Delhi.

Singh, A.L., 1992: Impact of Different Source of Irrigation on Cropping Pattern and Farm Practices, *Geographical Review of India*. Vol. 54, No. 1, Calcutta.

Singh, D.R., *et al.*, 2010: *Agro-climatic Atlas of Haryana,* HAU, Hisar.

Singh, H.P., *et al.*, 2001: Micro-Irrigation Control Board of Power and Irrigation, New Delhi.

Singh, J., 1974: *An Agricultural Atlas of India: A Geographical Analysis*, Vishal Pub., Kurukshetra.

Singh, J., 1976: *An Agricultural Geography of Haryana*, Vishal Pub., Kurukshetra.

Singh, J., and Dhillon, S.S., 1984: *Agricultural Geography,* Tata McGraw Hills, New Delhi.

Singh, J., 1977: A New Technique for Measuring Agricultural Efficiency in Haryana, *The Geographer*, Vol. XIX (1), Aligarh.

Singh, N.P. *et al.*, 2004: *Modern Technology on Vegetable Production*, International Book Distributing Co., Lucknow.

Singh, R.P., 1992: Concept of Land in Mohammad, N. (ed.) *New Dimensions in Agricultural Geography*, Concept Publishing Co., New Delhi.

Singh, R.P., 2007: "Role of Bio-fertilizer in Crop Production" in Sharma, A., Singh, R.K. (eds.), *Composite Farming Practice and Economic Development*, Abhijeet Pub., New Delhi.

Singh, T., 1998: Drought Disaster Management: A Indian Perspective, *National Geographical Journal of India*, Vol. 44, March-December, Nos. 1-4.

Sinha, R.R.K., 2009: *Dynamic of Land Caste Relationship,* Manak Publication, New Delhi.

Sivanappan, R.K., 1994: Prospects of Micro Irrigation In India, *Irrigation and Drainage System*, Vol. VIII, No. 1.

Sivanappan, R.K., 1987: *Sprinkler Irrigation*, Oxford and IBH Pub. Corp, New Delhi.

Srivastava, G.S., Singh, I.B. and Kulshrestha, 2006: Late Quarternary Geomorphic Evalution of Yamuna Sutlej Interfluves: Significance and Terminal Fan, *Journal of Indian Society of Remote Sensing.*

Sundaram, K., and Tendulkar, S.D., 1988: *"Measurement of Poverty: Some Conceptual Problems"*, Paper No. 36, Second Seminar on Social Statistics, February 4-6, 1988, Central Statistical Organisation, New Delhi.

Tarjuelo, I.M. *et al.*, 2000: Modelling Evaporation and Drift losses in Semi Arid Region, *Agricultural Water Management.*

Tendulkar, S.D., 1992: "Economic Growth and Poverty" in Harass, B., Guhan, S. and Csssen, R. (eds.) *Poverty in India: Research and Policy*, Oxford University Press, Bombay. I, Sections III and IV, pp. 32-42.

Tewari, R.C., 2006: *Geography of India*, Praying Pustak Bhawan, Allahabad.

Thakur, B., 2011: *Depletion and Degradation of Land Resource of India*, R.N. Dubey Foundation, Allahabad.

Tornquist, O., 1999: *Politics and Development: A Critical Introduction,* Uppsala University.

Trewatha, G.T., 1962: *The Earth's Problem Climate*, London.

Vaidyanathan, A., 2010: *Agricultural Growth in India: Role of Technology Incentives and Institutions,* Oxford University Press, New Delhi.

Vaidyanathan *et al.*, 1994: Impact of Irrigation on Productivity of Land, *Journal of Indian School of Political Science*, Vol. 6, No. 4.

Verma, K., 2007: *Role of Urban Agriculture in Life Support System, A*

Case Study of Alipur Block, Delhi (Unpublished M.Phil. Dissertation, Department of Geography, Delhi University, Delhi).

Verma, V.V.I. and Rao, A.R.G. (eds.) 1998: *Micro-Irrigation and Sprinkler Irrigation System*, Central Board of Irrigation and Power, New Delhi.

Wadia, D.N., 1960: *Geology of India*, London.

Wadia, D.N., and Auden, J.B., 1939: "Geology and Structure of Northern India", *Memories of the Geological Survey of India*, Vol. LXXIII, Delhi.

Weaver, J.C., 1954: "Crop-Combination Regions in the Middle West", *Geographical Review,* Vol. XLIV, No. 2, p. 176.

World Bank, 1997: *World Development Report 1997*, Washington D.C.

Index